5/29/98

ALLEN COUNTY PUBLIC LIBRARY
FORT WAYNE, INDIANA 46802

You may return this book to any agency, branch,
or bookmobile of the Allen County Public Library

DEMCO

ALSO BY MICHAEL WALDHOLZ

Genome: The Story of the Most Astonishing Scientific Adventure of Our Time—the Attempt to Map All the Genes in the Human Body (WITH JERRY E. BISHOP)

Michael Waldholz

SIMON & SCHUSTER

Curing
Cancer

The Story of the Men and Women Unlocking the Secrets of Our Deadliest Illness

Simon & Schuster
Rockefeller Center
1230 Avenue of the Americas
New York, NY 10020
Copyright © 1997 by Michael Waldholz
All rights reserved, including the right of reproduction in whole or in part in any form.
SIMON & SCHUSTER and colophon are registered trademarks of Simon & Schuster Inc.
Designed by Karolina Harris
Manufactured in the United States of America
10 9 8 7 6 5 4 3 2 1
Library of Congress Cataloging-in-Publication Data
Waldholtz, Michael, date.
 Curing cancer : the story of the men and women unlocking the
secrets of our deadliest illness / Michael Waldholz.
 p.   cm.
Includes bibliographical references and index.
 1. Cancer—Genetics—Research—United States.   2. Cancer—Popular
works.   I. Title.
RC268.4.W34   1997
616.99'4—dc21   97-24121 CIP
ISBN 0-684-81125-1

TO BETTY, RACHEL AND DANIEL,

FOR THEIR SUSTAINING LOVE

Contents

Preface   13

 1:   A Mystery Solved   15

 2:   Science Fiction   33

 3:   "Welcome to Chromosome 17"   42

 4:   Family 15   57

 5:   Trick or Treat   68

 6:   The Game   88

 7:   Cancer Families   95

 8:   Gold Mine   110

 9:   Guardian of the Genome   131

10:   Ishmael's Tale   148

11:   Clone by Phone   165

12:   The Mother of All Tumor Suppressors   184

13:   The Roller Coaster   211

14:   Ras   233

15:   First, Do No Harm   255

Epilogue   277

Acknowledgments   287

Notes   291

Bibliography   303

Index   305

Illustration Credits   319

"The human intellect has finally laid hold of cancer with a grip that may eventually extract the deadly secrets of the disease. . . . For the first time in my thirty years as a biomedical scientist I now believe that we will eventually cure cancer."

—J. MICHAEL BISHOP, NOBEL LAUREATE

Preface

A cure for all cancer, of course, does not yet exist. But for the first time in the century-long fight against the disease, scientists are producing the profound kind of insights that will lead to a cure.

As a veteran observer of scientific progress, mostly from my post as a reporter at *The Wall Street Journal* since 1980, I am well aware that it's often difficult to recognize a breakthrough unless it comes in a single dramatic moment. Except for a few instances, however, such as the discovery of the DNA double helix in 1953, or the first cloning of a mammal in 1997, most scientific advances are the result of accumulated pieces of research, uncovered bit by bit, in small-sized experiments and discoveries over the course of many years. The importance of each of these separate research steps often is missed or misunderstood. But, in fact, in laboratories across the country and around the world, scientists *are* uncovering the pieces of knowledge that will eventually produce cures for many forms of cancer.

Right now it may seem that we are still very far away from the day these scientific fragments finally come together. At this writing, more than 2.3 million Americans each year will be diagnosed with some form of cancer; more than 1,500 Americans will die each day of the disease; and 560,000 will die this year.

Yet, there is a lot of promising news, at least from the labs where basic science takes place. In just the past few years, researchers have uncovered genes that cause inherited forms of cancer of the breast and colon; and several dozen other cancer-causing genes have been identified or are on the verge of being revealed. The identification of these genes is giving researchers the first true picture of how cancer begins. And the same discoveries are already helping scientists to find new ways to detect and prevent cancer, as well as assisting them in their efforts to invent new and powerful therapies.

This book is the story of these discoveries and of the researchers whose collective work is giving us a new and dramatic understanding of cancer. The story told here is, in a sense, a work in progress, an attempt to recognize a breakthrough as it is happening. It is the story behind the cure, as it emerges.

1: A Mystery Solved

In 1993, sixteen years after his death, former Vice President Hubert H. Humphrey, or more accurately, a tiny part of him, participated in a remarkable experiment. Scientists in Baltimore, using new laboratory tools, solved a long-standing but little-known mystery surrounding the late politician's death. In doing so, the laboratory detectives uncovered information about Humphrey that might have changed the course of American politics in 1968.

But this experiment meant more than rewriting history; it revealed that science was in the midst of cracking open one of nature's deepest secrets: what causes cancer.

In their experiment, the scientific sleuths in Baltimore found the exact biological cause of the bladder cancer that killed Humphrey in 1978. The culprit was a tiny defect that somehow had taken hold in a single gene deep inside one of the numerous cells in Humphrey's bladder. The gene's job was to control the cell's growth. But because of the gene defect, which probably occurred early in Humphrey's life, the cell instead multiplied dangerously, over many years eventually expanding into a cell colony that is the hallmark of a cancerous and deadly tumor.

When Humphrey was diagnosed with cancer in 1971, and even when he died seven years later, nothing about this gene, not even the fact that it existed, was known. But in a series of insights in the 1980s and 1990s that turned cancer research upside down, scientists unmasked the identity of a handful of genes that work like master control switches, regulating the life cycle of a human cell. When these genes are somehow mutated, perhaps by some cancer-causing agent in the environment or by some internal mistake, otherwise healthy cells outgrow their neighbors, by turn becoming unruly and rebellious. Suddenly, university biologists plumbing the arcane genetic machinery inside cells were thrust from relative obscurity into the worldwide war against cancer.

There is now reason to believe that in revealing these master switches scientists have finally found the answer to one of the greatest medical mysteries of the century: what turns a normal cell cancerous. The numerous advances that led scientists to this brilliant deduction were the result of tens of thousands of experiments, some dating back decades, whose findings became pieces of a gigantic puzzle finally put together, beginning in the late 1980s, by a relative handful of researchers. As a result, scientists the world over suddenly realize they have within their grasp the ability to search for totally new ways to detect, treat, and perhaps even someday cure cancer.

The Humphrey experiment was simply among the more dramatic and noteworthy illustrations of this new power. Indeed, the Baltimore scientists who undertook the experiment initially believed that publicizing it might draw the world's attention to the fact that, largely unseen, a breakthrough in understanding the genesis of cancer had occurred. Moreover, the scientists believed that the experiment heralded another fact: something new and powerful could now be done about the disease that reigns as one of mankind's most terrible scourges.

Humphrey, of course, was an icon of liberal American politics following World War II. In 1964, Lyndon Baines Johnson plucked Humphrey from the U.S. Senate to be his vice presidential running mate. In 1968, after Johnson declined to run for reelection and Robert F. Kennedy was assassinated, Humphrey was chosen by the Democratic party to run against the Republican choice,

Richard M. Nixon. Humphrey, of course, lost to Nixon in one of the closest elections in American history.

But, unknown to Humphrey, his closest advisers, and the American electorate, a renowned physician at Johns Hopkins Hospital in Baltimore suspected that the candidate for president had a dangerous tumor growing inside him. The doctor, John D. (Jack) Frost, believed he held evidence of this extraordinary fact on a set of glass microscope slides which he had been sent to analyze. For two decades Frost stored the glass slides in his office desk drawer between two cardboard envelopes. The slides contained cells harvested from a urine sample given by Humphrey more than a year before he ran for the presidency.

In May of 1967, then Vice President Humphrey was concerned when he noticed several specks of blood in his urine. Within a few days, Humphrey's private physician quietly admitted the vice president to Bethesda Naval Hospital outside Washington, D.C. As part of the routine medical workup, the hospital's doctors took urine samples. Because of Humphrey's high office, several batches of the urine sample were placed on glass laboratory slides and shipped confidentially to several of the country's foremost cytopathologists, doctors who analyze individual cells under a microscope to determine by the way they look whether they are diseased.

Frost, director of the division of cytopathology at Johns Hopkins Hospital, believed some cells on his slides had undergone changes in shape and size characteristic of the very earliest stage of cancer. But most other experts apparently disagreed, and no treatment for cancer was recommended. Instead, doctors suggested that every six months Humphrey undergo a cystoscopy, a biopsy procedure in which a catheter is threaded through a patient's urinary tract into the bladder and cells are randomly extracted from the organ's inner walls.[1] Although it is unclear what Humphrey was told, he then vigorously pursued the political path that led him to his historic fight against Nixon.

Two years later, one of the regularly scheduled biopsy exams found that cells removed from the bladder lining had definitely undergone some slight alterations. What exactly this meant was unclear. The alterations suggested the cells were in an early stage of cancer, but there was no evidence that they had invaded the bladder wall or posed a serious danger to Humphrey. Today, a number of therapies are available to treat the problem, a type of tumor known as a *carcinoma in situ*.

Carcinomas of this type often are indolent, meaning they aren't a threat to grow massively or spread to other organs. But in 1969 the only way to excise the potentially dangerous tissue was through major surgery. Since the risk of surgery may have exceeded the danger posed by the carcinoma, no treatment was recommended for Humphrey. "The decision was made to just follow him," says Ralph H. Hruban, a pathologist at Hopkins drawn into the case later.

In 1973, another biopsy found evidence that the carcinoma had in fact perilously invaded the bladder wall, a strong hint it might be malignant. According to Humphrey's personal physician, Edgar Berman, the medical team on the case had eleven experts analyze the biopsied cells and give their opinions. Nine of them believed the tumor was benign but recommended radiation therapy anyway, one recommended doing nothing and one was so concerned that the tumor was already malignant that he recommended surgery to remove the entire bladder.[2]

Humphrey, who by this time had rejoined the U.S. Senate, was administered a course of radiation and chemotherapy at Memorial Sloan-Kettering Cancer Center in New York.[3] The treatment failed to halt the cancer. In August 1976, after the tumor had spread beyond the bladder, the cancerous organ was finally removed. But it was too late. In January 1978, at age 66, Humphrey died of cancer at his home in Minnesota.

About eight years later, Jack Frost, the Hopkins pathologist, was sitting in his office with Ralph Hruban, an intern at the time. "I was looking at some specimens under a microscope," Hruban recalls, "And [Frost] called me over to his desk and said, 'Let me show you something that'll wiggle your ears.' " From his middle desk drawer, Frost pulled out a worn cardboard

slide case identified simply on its cover with the words, "a patient of Dr. Berman."

Frost placed the slides under a microscope and told Hruban to
take a look. "What do you make of it?" Frost asked.

"They look a little like cancer," Hruban replied, noting that the
cells had degenerated, making the analysis difficult. "What am I
looking at?" Hruban asked, his curiosity piqued.

"Those are slides from the late Hubert H. Humphrey," Frost
said. "They were taken before he ran for president. I was certain
they were cancerous then. I'm certain now."

In 1991, in a tragic coincidence, Frost died of lung cancer, a disease he had spent his career studying. Before he died, he was careful to hand the cardboard folder containing the intriguing slides to
his wife, who passed them along to Frost's successor at the hospital, Yener S. Erozan. By then Hruban held a faculty position in the
same department as Erozan. One of Hruban's projects was working
with a small band of basic researchers at Johns Hopkins who were
attracting attention in the scientific community for exploring one
of the least understood regions of human biology: what transforms
a normal, healthy human cell into a deadly tumorous cell.

Beginning in the late 1980s under the direction of a previously
unheralded but brilliant investigator named Bert Vogelstein, the
Hopkins group made a series of stunning discoveries showing that
certain human genes inside the nucleus of a cell direct many as-
pects of cell growth. When these "growth-controlling genes" are
somehow damaged, the Hopkins team revealed, cells divide rebel-
liously, over many years gathering into an abnormal storm of tissue, known as a tumor.

Soon after Frost died, one young member of Vogelstein's team,
David Sidransky, published a landmark research paper. Sidransky
believed his colleagues' gene discoveries might make possible an
early-warning system for detecting a cancer. Sidransky thought
that thousands of lives could be saved every year by detecting a
cancer in its embryonic phase, long before it had grown large
enough to be seen by x-rays or other conventional diagnostic
techniques. When cancers are big enough to be seen by such tradi-

tional imaging methods, they often are already too vicious and too far along to cure. Sidransky set about searching for a way to identify in patients the genes that trigger the beginnings of colon cancer, the third most common cause of cancer death in the U.S., behind lung cancer and breast cancer.

Sidransky selected colon cancer because there was reason to believe that early-stage cancers shed cells into a person's stool, making it possible to collect and analyze cancerous cells without an invasive test. Using a complicated set of techniques pioneered in the laboratory of Bert Vogelstein, Sidransky compared the composition of growth-controlling genes in cells taken from a sample of a patient's tumor with genes in cells from the same patient's stool. To Sidransky's delight, he found that the same mutant gene in the patient's cancer could be detected in the patient's stool.[4] In other words, the stool from patients with cancer of the colon contains small amounts of mutant DNA from the cancer, making it possible for the mutant DNA to be detected.

To Sidransky, the experiment meant that as the genetics of cancer continued to be unraveled, sometime soon doctors would be able to screen a healthy person's stool for a genetic clue that an unseen and otherwise undetectable tumor was sprouting in a patient's colon. Forewarned, a patient could be closely monitored for the early appearance of a full-blown tumor, which could then be surgically excised long before it became lethal. The gene test, he believed, might someday be as easy as having one's cholesterol or blood pressure measured.

"David's paper was inspiring to those of us dedicated to finding new, better ways to diagnose cancer early," Hruban says. "Since most therapy for cancer still fails against advanced forms of the disease, the key to successful cancer management is early detection. But current techniques really are far from acceptable."

Sidransky's report was the first time anyone had shown that the gene alterations that kick a colon cell into a cancerous mode could be identified in a sophisticated but still relatively simple way. "The paper gave us all sorts of exciting ideas," Hruban says. "But it also got me thinking about Jack [Frost], who only recently had died, and his old Humphrey slides. Suddenly, it hit me. I realized there now might be a way to prove, absolutely, whether Frost was right or wrong. It was a bit farfetched but I decided to work on it."

To do that he needed to duplicate Sidransky's experiment, this time analyzing bladder (not colon) cells. He needed to see if any of Humphrey's ancient cells contained a cancer-causing defect in its genetic machinery. But unlike Sidransky, Hruban would have to carry out his investigation on cells from a patient who had died years before, and whose medical records, tissue samples, and consent were not going to be easily assembled. Hruban first made certain the old Humphrey slides still existed, eventually tracking them to Frost's successor. "Dr. Erozan had them sitting in his drawer, just like Frost did," he says. Before Hruban even received the necessary family consent to rummage about Humphrey's old cells, he called Sidransky.

"I was intrigued," Sidransky says. "It was the kind of challenge I like. But I told Ralph there was nothing we could do unless we had samples of cells from Humphrey's actual tumor." It was in cells from the cancer that the Hopkins researchers might be able to find which of the newly discovered growth-controlling genes triggered the cancer. The scientists first had to identify the precise spot in the gene that was altered in the tumor cells and then look at the urine cells to determine if the same site was mutated there, too. Only then could the scientists say with confidence that a gene mutation detectable in the urine cells was the one that caused cancer in Humphrey's bladder.

It is a regular practice in research hospitals to save a one-quarter inch slice of a patient's tumor, and store it in a block of paraffin wax in a pathology department vault. If Humphrey had been treated in such a hospital, a square of wax containing a bit of the bladder tumor still existed. But where? Hruban decided to contact Humphrey's widow, Muriel, since she certainly would know if and where her late husband's tumor had been surgically removed. He also knew he needed her permission to poke about the late politician's medical history and molecular makeup. Hruban wrote her explaining his idea and asking for permission to proceed. He was surprised by the quickness of her reply.

"She was very interested," Hruban says. "She said Humphrey had a philosophy that something good always comes from life's difficulties. But until I had contacted her, she said the family had never found anything good from Humphrey's death. 'Maybe,' she said, 'this will be it.' "

Fortunately, the late senator's tumor had been removed at Memorial Sloan-Kettering. A bit of wax holding the tumor slice was retrieved from the pathology department's storage bins. A month after he first contacted Sidransky, Hruban was back at the molecular biologist's lab with Frost's old slides and a sample of Humphrey's tumor. Sidransky decided to scan the tumor cells for a mutation in the same gene he'd found mutated in his colon cancer experiment. The Hopkins scientists recently had found that, as with colon cancer, the gene sometimes was altered in bladder tumors, too. But the gene can be disrupted, or mutated, in many different ways. If the gene is compared to a 100-mile-long roadway, for instance, any pothole along the way can be a cancer-causing disruption. In order to conduct the experiment, Sidransky needed to find the exact mutation, or pothole, that occurred in Humphrey's cells. And this wasn't going to be easy because there were hundreds of possibilities.

The Sidransky lab's ability even to consider probing the intricate architecture of a gene was the result of astonishing strides in molecular biology, the field of science that studies the structure and functions of genes. Indeed, the Hopkins molecular biology labs and several hundred like them around the world were making discoveries about the nature of genes at a pace so rapid it was difficult even for scientists involved in the field to grasp. Bert Vogelstein, Sidransky, and their crews were part of a gigantic effort by "gene hunters" around the world to map the human genome, the entire complement of 100,000 or so genes that underlie human biology. Often referred to as the Human Genome Project, this global enterprise has led many scientific historians to predict that the end of the twentieth century and the opening decades of the twenty-first century will be known in the future as "the age of the gene."

The dawn of this new age of discovery can be traced back decades before the Humphrey experiment. Every gene is composed of the remarkable chemical at the center of life called deoxyribonucleic acid, or DNA, a fact revealed in a series of investigations in the 1940s. The secret of DNA, or how genes can control cell division, eye color, or even one's sex, was uncovered in 1953 in the famous double helix discovery by British physicist Francis Crick and his young American colleague, biologist James Watson. They found that the structure of DNA, and therefore every gene, resembles a spiraling molecular ladder. They showed that the ladder's outer twisting rails, or strands, are linked at regular intervals by steps or rungs. Each rung of this ladder consists of two molecules called base nucleotides, or simply "bases," that hang off each strand and are bound together like two squares of wood nailed side-by-side to form the ladder's steps.

There are four kinds of molecules or bases, A (for the molecule adenine), T (thymine), G (guanine), and C (cytosine). But these bases are dotted along the DNA ladder by the hundreds of millions. Scientists now know all 100,000 human genes lie in a regimented order along three feet of tightly coiled DNA amazingly packed into the tiny nucleus less than 1/1,000 of an inch wide at the center of every human cell. A gene's individual structure is determined by the order in which the four bases are aligned; each gene contains an average of 1,000 to 3,000 paired bases. Thus, a gene's structure is ruled by the precise sequence of its several thousand "base pairs."

The base pair sequence, then, is a set of instructions, a recipe of sorts, that tells a cell how to assemble a protein, the substance that carries out the activities of life. Proteins, in turn, are long twisted necklaces made up of simple amino acid molecules. Each amino acid is described by three of the gene's base "letters." The letters CTG, for instance, tell a cell to build the amino acid leucine and place it in a specific spot in a forming protein, while the sequence GAT describes aspartic acid. A cell laces together a protein's numerous amino acid components by systematically "reading" the gene's three-letter words, making each gene a kind of "sentence."

These proteins, such as enzymes and hormones, carry out the moment-to-moment activities of life. Insulin is a protein that cells manufacture to convert glucose in the bloodstream into the energy needed to keep a cell functioning. If some of the base pair letters of the insulin gene are somehow shuffled or missing, the insulin gene will be defective or mutated. A cell containing a defective insulin gene may be unable to build a correctly working insulin protein. It would, therefore, be unable to process blood glucose and that would result in diabetes. In research accumulated in the late 1980s and early 1990s, scientists began to show that damaged or missing proteins resulting from mutated genes are at the root of many common diseases, causing such health problems as heart attacks, Alzheimer's disease, arthritis, depression and, of course, cancer, in addition to diabetes.

Researchers discovered in the 1980s certain mutated genes can promote unwanted cell division, which they named "oncogenes," employing the Greek word "onkos," which means "mass." Scientists also found that cancer can arise if just one of the millions of cells clustered in an anatomical region, such as the inner wall of the bladder, contains mutated oncogenes. That's because mutated oncogenes can generate unwanted cell divisions and over time the offspring, each containing the identical gene mutation, can pile up into a cancerous colony.

When Sidransky analyzed colon cells that fell off the tumor into the stool of cancer patients, he searched the tumor cells for misspelled DNA "words" in the oncogene that would have resulted in a mutant protein. He synthesized a bit of DNA containing the garbled letters and used it as a genetic probe, much like a piece of Velcro that was able to attach itself to the same altered letters or mutation in the cells taken from the stool sample. If the jumbled gene letters in the tumor and stool cells matched precisely, he knew that some of the sloughed-off stool cells had originated in a colon tumor. They were like a fingerprint left behind by a criminal.

Soon after Hruban received Humphrey's original tumor specimen from Sloan-Kettering, Sidransky assigned Peter van der Riet, a young researcher in his lab, the task of looking for a mutation of another gene, called p53. This gene also was known to control cell growth. Indeed, Vogelstein and his colleagues had dubbed it a "tumor suppressor," because in its normal unaltered form, the gene kept cells from becoming cancerous. Moreover, in a series of experiments destined to make Vogelstein famous, his lab found that tumors for many types of cancer contained damaged versions of p53. One of those studies showed that mutant p53 often was present in bladder tumors. By the 1990s, scientists had unraveled the entire sequence of the thousands of base pair letters that make up the important protein-producing region of the p53 gene. When van der Riet laid out the sequence of letters in the p53 genes found in Humphrey's tumor cells and compared the sequence with those found in normal p53 genes, he found his quarry. One of the many thousands of adenine (A) base letters somehow had been replaced by a thymine (T), an absurdly minute but apparently disastrous substitution. The subtle change made the p53 protein unable to carry out its job of controlling the cell division of bladder cells.

The lab then turned to the 26-year-old cells on the slides stowed away for so long by Frost and his successor. If they found the exact same p53 defect in the urine cells, it would mean they had found the mutation that caused Humphrey's cancer. "The immediate challenge before us was whether we could retrieve enough intact DNA from the old cells to conduct a study," Sidransky says. The solution lay with another new lab technique that allows researchers to produce a thousandfold increase in a bit of DNA so slight it would otherwise be impossible to analyze. This laboratory amplification is what inspired Michael Crichton's best-selling novel *Jurassic Park*. It was this same copying tool that Crichton's fictional scientists used to reproduce dinosaurs from DNA stored for billions of years in prehistoric bits of amber. The same technique also made it possible for prosecutors in the O.J. Simpson murder case to generate large amounts of genetic material from tiny blood samples at the crime scene in order to conduct DNA "fingerprinting."

Fortunately for the Baltimore investigators, Humphrey's cells retained their original chemical composition. Once the cells were scraped off the slides, mixed in a variety of solutions, and then passed through the copying process, van der Riet was able to conduct his experiment. Within a few weeks of analysis, the lab found the telltale genetic fingerprint previously identified in cells extracted from Humphrey's bladder tumor. The match, the scientists agreed, was so precise that no other explanation existed: the cells on Frost's slides and the cells from the tumor embedded in the paraffin block had originated from the same deranged progenitor cells that had been growing in Humphrey's bladder in 1967. Frost had been right. It had taken a revolution in biology to prove it.

"We were bowled over," says Hruban. "Here at Hopkins we are well aware of the power of our gene discoveries and their potential for radically changing the diagnosis and treatment of cancer. But, pulling this [experiment] off required a great deal of luck, the exacting use of some very new technology, and an unusual commitment from the lab."

In a seminal research report published in the *New England Journal of Medicine*, the Hopkins team proclaimed their finding. "Remarkably, the cells harboring p53 mutations in his urine were detectable nine years before Humphrey underwent [surgery] . . . and six years before he received any therapy for bladder disease."[5]

Moreover, the mutation of p53 told the researchers something else. New studies emanating from the Vogelstein labs strongly suggested that cancer cells with a p53 mutation were more likely to be aggressive, invading nearby tissue and spreading dangerously through the body. "[Mutant] p53 is a very bad character," Sidransky says. "We now know that even very small tumors containing a p53 mutant have very poor outcomes. When we see mutant p53 in an early stage cancer we know we must treat the cancer with very aggressive therapy. That's very powerful and important information for a patient and the treating doctor."

The p53 gene defect "raises the possibility," Sidransky and Hruban argued in their research paper, that Humphrey's cancer "was already in a phase of aggressive growth in 1967. Had

Humphrey known that he had aggressive bladder cancer in 1967, he might have withdrawn from the presidential race."[6]

The news shook doctors who treated Humphrey. Dr. Willett Whitmore, who removed Humphrey's cancerous bladder at Memorial Sloan-Kettering in 1976, says Humphrey rejected his recommendation to have the surgery three years earlier when a biopsy revealed cancer spreading in his urinary tract. "If we knew that p53 was a factor, we could have twisted his arm," Whitmore says.[7]

There is evidence that if Humphrey knew the danger he was in, he would have acted differently. After his disease became apparent, Humphrey told a biographer: "I would have resigned the nomination immediately [if] I knew of my illness."[8]

But, for the practice of medicine, the Humphrey experiment had even broader implications. "The experiment showed definitively that we can pick up cancer years before the tumor is picked up by normal clinical methods," Dr. Sidransky says. "That's the real reason I was interested in conducting the experiment. The discovery that certain identifiable genes trigger cancer means that we now have in our hands the ability to screen people without symptoms and detect a cancer we wouldn't know existed. Patients whose cancers are found in an early, localized stage have a ninety percent chance of surviving at least five years. But if the cancer is detected after it has spread, the five-year survival rate drops to nine percent."

"We hoped by looking at Humphrey's cancer we would produce a dramatic way to publicize the power of this new technology," Sidransky says. "It's our belief that the public, including many doctors and hospitals, are unaware of the potential of these new [gene] findings, and that they soon can be harnessed to save the lives of thousands, perhaps millions, of people." When the experiment was conducted in 1993, Sidransky's lab had yet to develop techniques that would allow doctors to make routine use of such a test. But Sidransky and his colleagues at Johns Hopkins boldly proclaimed that soon—perhaps by the end of the century—a new era of "predictive medicine" in cancer would make gene tests as commonplace as biopsies, CT scans, MRIs and x-rays.

The Baltimore scientists are not alone in their enthusiasm. In the past several years, many of the world's most eminent basic scientists have come to believe there is no more dramatic, exciting, or important story in medicine than the hunt for the human genes, such as p53, that underlie cancer. There is reason to believe that in uncovering these genes scientists finally have found the key to cancer. Indeed, they already have used the key to open a new door into cancer research, one that many scientists believe will lead them someday to cures.

In 1993 few people outside the biomedical science community were aware of the cancer gene story or its implications. Nonetheless, what is taking place in gene-hunting labs is truly momentous, involving a technology that may surpass atomic power or computers in its eventual effect on human life. But the cancer gene pursuit is more than great science. It is a compelling human saga involving a handful of dedicated scientists competing against one another to make tremendous advances, accompanied by immeasurable prestige and glory. There are stories of families tragically drawn into the annals of science because they are strangely haunted by deadly and mysterious illness common among their relatives.

Scientists now believe the genes that they are studying normally act as critical control switches deep within the core of every human cell, regulating the fragile, complex cycles of a cell's life by signaling the cell when to divide and when to stop dividing. Scientists are convinced that if these genes somehow become defective, normally well-behaved cells divide without check, eventually breaking loose and invading other tissues and organs.

This genetic scenario provides an entirely new and surprisingly simple model for understanding cancer, the most common, deadliest and baffling of diseases, killing more than one half million people in the United States each year. This scenario may prove to be the universal cause of cancer long sought by scientists. As the Humphrey experiment shows, these growth-controlling or tumor-suppressing genes already are providing novel diagnostic tools to accurately spot the presence of cancerous tumors so tiny that they previously evaded detection.

Until now, scientists were perplexed as to how and why a cancer arises. Scientists and doctors long have known that certain substances in the environment, such as toxic chemicals, gases, radiation or even certain kinds of diet, can be cancer-causing. But researchers are stumped as to why some people get cancer as a result of such exposure, while others don't. Why, for instance, will one longtime cigarette smoker develop lung cancer, while another won't?

At the same time, cancer doctors long have suspected, without proof, that a tendency to cancer can be inherited, passed along from parent to child. How else, they wondered, can the tragic clustering of cancers in some families be explained? Scientists guessed that an individual's susceptibility to, or defense against, cancer was the result of differences in genetic makeup, in a person's so-called inherited constitution. But scientists couldn't prove this idea because they had no way to identify the specific gene or set of genes that might vest someone with a vulnerability to cancer.

Now, the gene-hunting researchers say, they have the answer.

As a result of the recent gene discoveries, the scientists can now predict in some families who is born with an especially high risk of developing cancer and who is not. This feat has produced extraordinary human dramas. In 1994, in what scientists are hailing as one of the greatest gene discoveries so far, scientists in Utah found a gene that causes an inherited form of breast cancer. Researchers say that this gene, and others like it, is responsible for 7 to 10 percent of all breast cancer cases, perhaps affecting about 1 in 200 women. If true, that means that breast cancer is the most common lethal inherited disease yet uncovered, affecting many more people than such genetic diseases as cystic fibrosis, muscular dystrophy, hemophilia, or Down's syndrome.

The breast cancer gene research resulted from a tedious 20-year international search for large families where breast cancer was unusually common. Immediately, women in many of the research families were able to make use of the discovery. One woman in Great Britain was tracked down by British scientists. The scientists told the woman she was part of an extended-family whose members had cancer. A DNA test found that she very likely carried a defective form of the gene. Several days later she had a mammography that located the tiny beginnings of a tumor, which was removed. Her doctors believe her prospects for full recovery are bright. "She had no reason to suspect she was at risk until we told her and she probably wouldn't have detected the tumor until it had grown dangerously large," her doctor says. He adds: "We're still pretty stunned by what's happened. It's as if we had a crystal ball and predicted the future."

Sometime in the future, scientists say, this kind of predictive cancer testing will be as common as tests that identify a person's risk of heart disease by measuring cholesterol. But scientists expect the gene findings to propel research even further, helping to create new, safer and more potent therapies designed to stop a cancer by literally turning off a tumor's malignant motor, perhaps by repairing or replacing the defective gene that causes the cancer. Already scientists have conducted laboratory tests in which they halted tumor growth by implanting a healthy version of a cancer gene into fast-dividing cancer cells. Several other experiments suggest that such "gene therapy" might reverse a cancer's course, shrinking potentially dangerous tumors. This represents a radical departure, for it embodies the kind of rational, directed treatment that cancer doctors long dreamed about.

Today, in addition to surgically removing a tumor, doctors battle cancer by bombarding cancer cells with toxic drugs or radiation, thus poisoning the cells to death. These noxious therapies seriously harm healthy tissue and make cancer patients severely ill. Worse, these therapies often fail to erase the cancer. The new cancer gene discoveries may change all that. By understanding the mechanism of a cancer cell at the genetic level, at the most basic inner workings of a cell, scientists are convinced they can tailor new medicines to interdict a cancer growth with precision and power unknown before. They argue, quite convincingly, that new gene-based therapies will someday make current toxic cancer treatments appear primitive, akin to the medical use of leeches or incantations in centuries past. Even the most cautious of gene hunters believe they finally are on the trail of crucial clues that may make it possible to conquer cancer.

Doctors aware of the quest for cancer genes say that even if cures are unavailing soon, the discoveries will produce powerful new ways to prevent the disease from occurring in the first place. The cancer gene theory suggests that each of us is born with a genetic profile that, once detected, can be used to predict our individual risk of carrying the seeds of our own destruction. Those found to be at unusually high risk may find they can dodge the cancer bullet by radically changing how they live, avoiding exposure to those things that can damage the genes, that cause the gene to become mutated.

But the power of this predictive information is also troublesome, because knowing about one's cancer fate without any cure can leave patients in a terrible quandary. Doctors and medical centers will need to create entirely new ways to counsel and treat patients and their relatives. Already, the tests being used in some research families are causing turmoil for health officials and scientists perplexed by what to do with this new knowledge. At what age, they wonder, should young girls in these breast cancer families be tested and notified of their cancer risk? Should scientists advise family members to keep their gene profiles hidden from health and life insurers, from employers, friends and relatives who may not understand the nature of the gene's power and the limitations of a gene test? Should the gene tests eventually be available for use by the general public? If so, who will pay for such testing and for the emotional and medical counseling that should accompany negative findings?

Certainly, whether they want it or are prepared for it, the day will soon come when insurers and employers will be flooded with new information about beneficiaries' and employees' cancer risks. Personal behavior and environmental and occupational health policy will be profoundly affected, as scientists use the gene discoveries to finally identify the precise external forces in our lives that trigger cancer by damaging certain genes.

To understand how far scientists have traveled toward this goal readers need only follow the cancer gene quest chronicled in the pages that follow. But this story is not only about a breakthrough in cancer. It is also about the nature of scientific inquiry in the closing days of the twentieth century and the people who conduct it. This particular scientific saga involves numerous people, but our portion of the tale will follow several actors whose roles are central to the narrative. They include a young feminist who dedicated almost twenty years to finding the essential clue to the most common and deadly of diseases among women; a mop-haired, born-again Christian whose competitive fire burns as intensely as his religious values; a computer whiz who bucked the scientific establishment with an unorthodox venture in order to fund his quest; and an ascetic Jew whose high school nickname was "Rock" and who, colleagues bet, will someday win the Nobel Prize. But for our purposes, the narrative begins with two sisters at a hospital in Michigan where a scientific discovery changed their lives forever.

2: Science Fiction

Driving past thick cornfields and golden pastures on a bright August morning in 1992, Susan Vance and her two companions spoke little as they headed toward the University of Michigan's sprawling medical center in Ann Arbor. Susan was planning some extraordinary action, and her husband, Bill, and older sister, Janet, didn't want to let her do it alone.

"I want my breasts off," Susan told her relatives. "I want it done as soon as possible." No one tried to change her mind. The long trip by car to the university's hospital that day was to finalize details. She planned to undergo the surgery the following week even though she showed no signs of disease or illness. Just the opposite was true: she seemed in perfect health.

Susan, Bill, and Janet were too exhausted by the previous weeks of deliberations to argue about their options anymore. Susan and Bill were in agreement about the decision. Only Janet held out hope that something might yet happen to keep Susan from taking such desperate action. But Janet kept her thoughts to herself.[1]

Susan, a robust and attractive 36-year-old mother of two, was thinking about how she had spent years trying to escape the terrible truth stalking her family. Since 1978, her mother, her mother's sister, two first cousins, and her two sisters, all of whom lived within a few miles of one another, were struck by breast cancer.

Three of the women, all less than 45 years old, had died from cancer. Because breast cancer is so common, it isn't that strange for more than one woman in a family to fall victim to the disease. National statistics bear this out. One in eight American women will develop breast cancer; specialists predict that it will strike about 16 million living Americans. More than 185,000 new cases arise in the United States each year. And the cancer accounts for 45,000 deaths annually, making it the second-largest annual cause of death among American women after lung cancer.

For years, Susan and the men and women of her large family, descendants of a German immigrant farmer who settled in the region more than a century earlier, believed they were the victims of either some very bad luck or the common exposure to some terrible, but unknown, cancer-causing substance in their water, air, or food. In just eighteen months, between February 7, 1988, and December 23, 1989, two of Susan's first cousins and her older sister, Lucy, died after protracted battles with the disease. Each of the women left behind a husband and several young children. Then, in June of 1992, while the family was still in shock and mourning over the young women's premature deaths, Susan's other sister, 41-year-old Janet, found out she too had breast cancer. Susan reluctantly conceded that neither chance nor the environment had anything to do with the disease. She now was certain that some inborn trait was being invisibly passed between generations of the family's women.

For more than a year following the three young women's deaths, Susan and Janet sought a way out, traveling to the famed Mayo Clinic, to the University of Michigan and to other medical centers, hoping one of the doctors could determine if they were destined to share the fate of the other women. They hoped that if they searched hard they could track down an experimental technique to identify the earliest signs of disease before it became deadly. But then one night as she lay quietly in bed, Janet found a lump. Doctors told her the cancer was small. But it had progressed to a dangerous malignant phase even though it had evaded detection just a few months earlier when Janet underwent one of her routinely scheduled breast examinations.

Susan became desperate. She had trouble sleeping at night, haunted by recurring memories of her sister Lucy's four-year fight against the disease's relentless advance. "I wanted to support Janet, to tell her she'd be all right," Susan recalls. "But I couldn't. We'd all seen what this disease can do, we'd lived through one nightmare after another. When I saw Janet I just started crying hysterically. I fell apart. The whole family fell apart. I kept thinking over and over, 'How can God do this to us? What did we do to deserve this?' "

Susan and Bill finally agreed they had little left to decide but the specifics of the surgery. Trim and athletic, Susan wanted reconstructive surgery that promised to restore some measure of her full-busted figure. Equally important, she believed she had to act swiftly to protect herself and to spare the family the incalculable pain of one more life cut short.

But then something remarkable happened. As Susan spoke with surgeons at the university's hospital later that morning, researchers elsewhere in the medical center who heard of Susan's plans quickly huddled together to discuss what to do with an incredible piece of information they had just uncovered about her.

"We were surprised to hear that Susan was in the building and why," says Barbara Weber, back then a cancer physician and genetics researcher at the university. Weber was part of a team of scientists poring through DNA from several dozen families including Susan's in pursuit of a gene that might predispose women to breast cancer. Only a few days earlier, results had emerged from the university's molecular biology laboratories showing that the cluster of breast cancers in Susan's family was due, not to chance, but to a shared defective gene. When the researchers analyzed the laboratory results comparing the genetic makeup of people in the family, they believed that they could tell from a relatively simple blood test who had inherited the still unidentified gene and who hadn't.

"Our sole interest was to use the family's DNA to find the gene itself," says Dr. Francis Collins, who at the time headed genetics research at the university. Collins and others believed that identifying the gene would open previously hidden biochemical pathways that ruled the disease. This development, they hoped, would lead to new ways to diagnose the disease early when a tumor is tiny and most successfully attacked with radiation or drugs. Maybe, they hoped, the gene would even help them plot out entirely new research avenues that someday might lead to more potent anti-cancer medicines. "We had given practically no thought to using the information we had about the family for any other purpose than to track down the gene and then use it to conquer the disease," Collins says. "Then, suddenly, we realized how naive we'd been. Looking back I feel quite foolish. You see, in the matter of a few moments we realized we had information in our hands that could dramatically change people's lives, right now, not sometime in the hopeful future."

Advised that Susan was at the hospital and planned to go back home shortly, Weber's head began to spin. "I instinctively knew what I had to do," Weber says. "But as far as I knew nobody before ever faced the same dilemma. There weren't any rules to follow, at least none that I knew of or could consult at a moment's notice."

By a coincidence that later would make Weber believe a spiritual force was at work, Janet had seen Dr. Weber only a few days earlier. After learning she had cancer back in June, Janet had sought treatment at the university hospital and, simply by chance, was referred to the oncologist on call, Barb Weber. During her initial consultation, Weber realized that Janet was a member of a large family recruited by Weber's research colleagues. Weber played a dual role at the university not uncommon at medical research centers. She treated patients and also ran a research lab, where she oversaw projects in biology and helped train graduate and postgraduate biology students.

Within weeks of the cancer's detection, Janet had a surgeon perform a radical mastectomy, removing the right breast where the small lump had been found, as well as her left breast, although there was no evidence of a tumor there. Once cancer was found Janet was frightened by her breasts. "I didn't want them anymore," she says. "I no longer even wanted to try to save them. My attitude about them changed completely. I didn't even want to have plastic surgery, reconstruction, or implants. I didn't want anything to do with them."

After the surgery, Janet was referred to the University of Michigan's oncology department for additional care. At the first visit, Weber scheduled for Janet a monthly regimen of chemotherapy and injections of several anticancer drugs that were administered at the university hospital. Four days before she, Susan, and Bill drove down to Ann Arbor, Janet had seen Weber as part of her monthly checkup and chemotherapy administration. The day after that, Janet called Weber as she was concerned about a side effect she was experiencing. "I would call Barb all the time," Janet says. "Everything scared me. Every little thing bothered me. I was certain I was facing death just like my sister and cousins, and I wanted to make certain we were doing everything we could."

During the phone conversation, Weber inadvertently let slip a piece of intelligence she had meant to keep secret. She passed it along in such an offhanded way that later Weber would have trouble recalling details of the phone conversation. Weber was bursting with news she instinctively knew might provide a bit of a salve to Janet's anguish. "We can tell whether your daughter inherited the same risk of breast cancer that you have," Weber said.

As she sat on her daughter's bed in the compact ranch-style home she shared with her husband and two children, Janet wasn't certain she understood what she'd just been told over the phone. During the previous year, the University of Michigan researchers had given the family only cryptic bits of information about the women's role in the hunt for a gene. Janet recalls being told that whatever emerged from the research wasn't likely to help anyone for years, maybe for many years. What did Weber mean? In a brief conversation that followed, Janet absorbed enough to know that the scientists were able to track a gene they believed responsible for the breast cancer in the family. They could tell whether Janet's 18-year-old daughter, Meg, bore that gene. "Finally," Janet thought, "God has answered our prayers."

Janet hung up the phone and walked a few steps down the hall to the small room her husband, Tom, used as an office for his real estate business. After explaining her brief talk with Weber as best she could, Janet told Tom, "If they can tell us about Meg, they probably can tell us about Susan, too." Later that day Tom called Bill and told him what Janet had learned. But Bill decided not to pass the information to Susan. It had taken him weeks to persuade Susan to agree to undergo a double mastectomy. Unclear exactly what it meant, Bill didn't want Susan to hear anything that would give her reason to waive or postpone the surgery. Janet believed she shouldn't intervene. So on Friday, she traveled with Susan and Bill to the university as planned, determined to keep quiet about Weber's news.

Later that morning, the three sat in the plastic surgeon's office about to look at photographs of women who had undergone the type of breast reconstruction scheduled for Susan the following week. Janet excused herself and left. "I couldn't get what Barb had said out of my mind," Janet says. "I wanted to know what it all meant." A few minutes later she was at the receptionist's desk in Weber's outer office a few floors away. "I want to talk to Dr. Weber," Janet said. "It's important."

That's when Janet advised Weber that Susan was in the building, and that she was completing plans to have both her breasts surgically removed. Weber remembers feeling a chill. "We had only completed a linkage analysis the week before," Weber says. "So someone down in the lab probably knew who was linked to the gene and who wasn't." Weber told Janet to wait where she was. Back in her office she began frantically punching phone numbers. First she called the lab, and had a colleague there read her the data from the linkage analysis. Then, with the information confirmed, she called Francis Collins.

Collins is one of the premier scientists in the field of gene mapping and discovery. In 1989, he burst onto the scientific stage when his lab at the university helped isolate the gene that causes cystic fibrosis, a relatively common inherited disorder that often kills its victims before they reach adulthood. By "cloning," or identifying, the gene, Collins beat out a half dozen of the most famous molecular biology laboratories in the United States and abroad that had been chasing after the gene for more than five years. Moreover, he snared the gene by using a new set of techniques devised in his lab called "positional cloning." This was a tedious but creative combination of experiments that allowed researchers to fish out of the jumble of DNA a tiny stretch of genetic material believed to contain the gene and then slowly narrow down the region under investigation until the gene itself was revealed. Collins's feat brought him to the apex of the small gene-hunting community. Rewarded with funding from the National Institutes of Health, the vastly rich Howard Hughes Medical Institutions, and the University of Michigan, Collins quickly built around him a complex of genetics research labs, staffed with bright and creative molecular biologists. Under his direction, the laboratories set as their target some of the biggest gene game there was, the genes underlying a wide range of diseases, including the gene believed to trigger familial breast cancer.

Francis knows what I should do, Weber thought as she tried to phone him. But Collins was away and couldn't be reached until later that day. Weber was on her own. "After about five or ten minutes of trying to get Francis, I decided I didn't want to hear his opinion anyway," Weber says. She worried he might order her not to do what she now believed was right. She told Janet to bring Susan and her husband to her office.

Back at the surgeon's office Susan was in no mood to see another doctor. A clerk inadvertently had rescheduled her surgery for a later date, and Susan had spent much of the time Janet was gone angrily getting it put back to its original time, on the following Friday. Janet quickly told Susan that Weber wanted to see her and Bill. "I think she has some important information you should hear before you do this," Janet said.

In the brief time before Janet returned with her sister, Weber collected her thoughts. It really was a fluke that this was happening at all. Few molecular biologists interact so intimately with research subjects. Usually, the gene scientists collaborate with a physician, perhaps located hundreds or thousands of miles away, who has uncovered a family where a disease is uncommonly persistent. Often the only connection the lab scientist has with people afflicted by a disease with a familial pattern is by combing through their DNA extracted from blood samples shipped to the lab by the family physician. Information generated by studying the DNA is rarely passed back to the research family. It was only by chance that Janet had been referred to Weber and not some other oncologist on staff. Yes, she thought as her nurse led Susan, Bill, and Janet into the examining office, this is truly a unique situation.

Weber's organized and confident manner had persuaded Susan and Janet weeks before that if there were a way out of their predicament, she would find it. But now the doctor spoke quickly and nervously, trying to describe what had happened without being too technical or getting ahead of herself. She explained that for two years scientists in laboratories around the United States had been hunting for a specific gene believed to cause inherited breast cancer in many families. She told them scientists in California had isolated a "marker" for the defective gene, a tiny fragment of DNA that wasn't the gene itself, but was always inherited along with it in these families. Studies of numerous families like Susan's had persuaded the scientists that women who carried the defective gene had an 85 percent chance of developing breast cancer and that they likely would be stricken early in life.

"We now know why your family has so much breast cancer," Weber told them. She said in the past few days the Michigan research team had become certain that DNA from Susan's mother and aunt contained the telltale gene marker. So did DNA from Susan's deceased sister, Lucy. The same was true of DNA from the two cousins who had died from breast cancer. And, yes, Weber said, Janet carried the same gene, too. Everyone in the family who had breast cancer, and many who hadn't yet developed the disease, carried the gene, Weber said. But female relatives born without the defective version of the gene still have a risk of getting the cancer. But it was the same one in eight risk of developing breast cancer that is faced by women in general.

Weber ended her brief lecture, hoping the family understood.

"You're saying you can tell me whether I have the gene?" asked Susan slowly, barely able to speak.

The room was quiet. Susan inhaled and exhaled quickly. "This isn't happening to me," Susan thought. "Yes I want to hear the results," she told Weber, thinking at the same time, "This is science fiction. Stuff like this happens on TV, not to me."

3: "Welcome to Chromosome 17"

TWO years earlier, in October 1990, Mary-Claire King strode literally onto the scientific stage with a startling announcement. Speaking from the podium of a convention center ballroom in downtown Cincinnati, King riveted a late-evening audience of top scientists gathered as part of an annual week-long meeting on genetics. King said her unheralded small laboratory in California had uncovered what many believed was a "Holy Grail" of modern medical science: the first persuasive evidence that breast cancer can be caused by inheriting a single mutated gene. No one had ever made such a claim before.

At the time, researchers in the young field of human gene discovery were dropping scientific bombshells with a regularity astonishing even to the scientists themselves. Each week brought news that scientists had identified the existence of a previously unknown or hidden gene. Mapping the entire genome—that is, identifying all 100,000 or so human genes—by early in the twenty-first century was no longer the farfetched goal of a small band of scientific visionaries. Indeed, a sign that the science of gene-hunting had advanced significantly came earlier that year when the U.S. Congress created a federal agency to allocate funds and oversee the gigantic gene-mapping effort.

But none of the findings to date matched the potential public health impact of the claim put forth that night by King. Even in this frenzy of futuristic findings, King's report was universally hailed. Within a few days news of her discovery reverberated throughout all of medicine, cancer research and human biology. Within a few months, King's 20-minute talk was scientific lore.

"It was one of the most exciting pieces of basic research we had heard in a while," says Anne Bowcock, a young geneticist at University of Texas Southwestern Medical Center in Dallas. "I'd say we were electrified. It meant there might be a way, finally, to solve the underlying root causes of breast cancer. And if it was possible to do this for breast cancer, then it meant [King's lab] had shown the way to solve the other cancers, too."

By the time genetic scientists had gathered in Cincinnati for the annual get-together of the American Society for Human Genetics to present and discuss research advances of the past year, scientific investigators in labs the world over were tracking down genes that play underlying roles in some of the most vexing public health problems. These new gene discoveries provided medical researchers with entirely new ways to attack such intractable illnesses as heart disease; mental health disorders such as schizophrenia, Alzheimer's disease, alcoholism, and depression; allergies; arthritis; diabetes; and, of course, the killer cancer.

Still, only a few moments before King addressed the research crowd, she fretted over the details of her talk. In a last-minute surprise move, the meeting's organizers had squeezed King onto a panel reserved months earlier for scientists who had been blazing a trail in the brand new field of cancer genes. But just weeks before the meeting, King's laboratory, staffed with a handful of young graduate students at the University of California at Berkeley, quietly made a breakthrough that ended King's fifteen-year research crusade to prove her conviction that women can inherit their breast cancer. King decided to break the news of her lab's success in public before her peers, many of whom were her staunchest critics.

Leaning against the back wall of the meeting room waiting her turn to talk, King worried she might be wrong. It wouldn't be the first time. Over the years, her insistence that a breast cancer gene existed at all was met with unyielding skepticism. Several earlier hints of the gene's possible identity uncovered by her lab had led nowhere. There was even talk among some scientists that King had fabricated previous findings. "I was under enormous attack" for research claims she made in the mid-1980s, King recalls. All of which had only fueled her resolve and near-obsessive zeal to prove she was right.

At 44, Mary-Claire King was very much a product of the activist 1960s, when young adults believed they should dedicate their life's work to improving people's lives. After receiving a graduate degree in genetics, King worked briefly for Ralph Nader's consumer advocacy organization. As a teacher and enthusiastic supporter of progressive issues, King had transcended the typical portrait of the laboratory scientist insulated from the people directly affected by scientific advances. A committed feminist, King often said her life's goal—indeed, her personal obsession—was to find a completely new way to battle breast cancer.

To her students and colleagues, King's single-minded, stubborn search for the breast cancer gene raised her above those in the field whose lust for discovery often seemed driven as much by a desire for glory and personal gain. "M-C is a hero for an entire generation of young women scientists," says Lori Friedman, a former graduate student in King's lab.

As she prepared to address the gathering in Cincinnati, however, King battled the insecurity she had learned to hide from both her defenders and detractors. To calm her nerves and fortify her convictions, King chose to stay off the stage until her turn to talk. Instead, she stood at the back of the meeting room next to her scientific mentor and longtime cheerleader, Luca Cavalli-Sforza, an eminence grise of the international world of genetics. "I wanted Luca to assure me that I wasn't out of my mind," King says. "Luca told me he had been skeptical until I showed him my statistical data."

"Show them what you showed me," he said, giving his younger colleague a gentle hug.

It was past 9 P.M. and she was still plenty nervous as she walked to the podium. "A lot of very smart people were convinced it was simply impossible to find a single gene [for breast cancer] with the current state of technology," King says. "Some of them had even written that my approach was a waste of time. I knew the data were going to be put under very close scrutiny, that many people would think that our finding was a fluke, a product of chance, not truth."

Up on a theater-sized movie screen, King presented the audience a series of slides containing charts, graphs, and family trees whose branches stretched up and through several generations and dozens of relatives. In recent years, she said, her research team had collected information about twenty-three unusually large families from around the United States, Puerto Rico, Canada, the United Kingdom, and Colombia whose female members seemed to be under siege. With a frightening regularity that followed the pattern of an inherited disease, nearly half the adult women in these families developed breast cancer.

Moreover, much of the cancer hit before the women reached menopause, a striking trait because most victims aren't diagnosed with breast cancer until after age 50 or so. This cancer, King said, was most likely a result of some inborn biological error, probably a mutated gene, that thrusts a woman onto a premature path to cancer very early in life. Women without this gene defect develop their cancers later, King argued, because cellular changes that lead to disease don't arise until many years have passed.

Previous attempts to isolate a single genetic culprit had failed because scientists were unable to distinguish which cancers were caused by inheriting a gene and which were among the vast majority that arose without any connection to a person's family history. In her report, King said her lab had finally solved this vexing problem. A year earlier, they decided to seek out the gene by probing DNA from families where the early-onset type of cancer was prevalent.

"One day a light went on in our lab," King says. "We realized that the gene, if it existed at all, was likely to be found by focusing most especially in families where cancers struck women by the time they were 45 years old. That's where to look for the gene."

After months of sifting through the DNA taken from such women, the King lab finally hit pay dirt. One tiny region of genetic material residing on the 17th of the 23 human chromosomes was consistently altered in DNA from women with cancer. In women without disease, there was no abnormality. Moreover, the statistical link between carrying this abnormal length of DNA and getting the cancer was overpowering, many times stronger than any other association that King or anyone else had yet found. "Our lab has made a preliminary determination that a gene for inherited susceptibility to breast cancer is located on chromsome 17," King told the gathering. It was hiding, however, in a giant region, one that was about 20 million base pairs long, akin to "pinpointing" a needle in one farmer's haystack.

"The data suggested the presence of a single gene," King recalls. "But I wanted to treat my own data with skepticism. I said the finding was preliminary and that it had to be reproduced by another lab. I thought we'd located a gene, I really did. But was I one hundred percent certain? No. Absolutely not."

Those listening to the presentation were impressed. "It looked like the real thing to me," says Louise Strong, a genetics researcher at M. D. Anderson Cancer Center in Houston who helped organize the meeting. "Only a few of us at the meeting knew that Mary-Claire was going to announce that she had mapped the gene," says Strong, who, despite some reservations, had agreed to place King on the program as last speaker of the evening. "I can tell you there was a great deal of healthy skepticism simply because Mary-Claire had published a finding years before that had not held up. But her data that night were pretty convincing. I know some people came to the meeting ready to attack whatever Mary-Claire had to say. The reaction was pretty civilized."

In addition to the mapping claim, King generated additional controversy in her talk by asserting that the gene she had found might be carried in as many as 1 in 200 women. If true, it meant that familial breast cancer suddenly was among the most common of all the 4,000 or so known "inherited diseases," even more common than cystic fibrosis, muscular dystrophy, or any other disorder resulting from an inborn defect.

Among those enthralled by King's talk was Francis Collins. "I was astonished, maybe because it came as such a surprise," Collins says. "I didn't even know what she was going to talk about until she began speaking."

A tall lanky man with a mop of blond hair and preference for rugby shirts, black jeans and sneakers, Collins often looks more like an earnest high school biology teacher than what he is, one of America's premier gene cloners. Most recently, his research team had won public acclaim for discovering the inherited gene that causes neurofibromatosis, a disfiguring illness often identified inaccurately as the elephant man's disease. Coincidentally, that gene was found lurking on chromosome 17.

As King spoke, Collins could barely sit still. He had never met King before but right away he decided to confront her alone after the session ended. He had an offer he knew she was in no position to reject. He already had spent years exploring the genetic territory within the boundaries of chromosome 17. Mary-Claire King had just declared that a breast cancer gene existed, and her lab had pointed to its approximate whereabouts in the human gene map. But its precise location needed to be uncovered. Accomplishing that feat would require the technical prowess of the world's very best gene explorers. Nobody in 1990 fit that description better than Collins, and he knew it. In that moment Collins decided he had found a new mark for his gene-hunting technology: the first human gene ever known to cause breast cancer. It was a trophy certain to seal his reputation. After a crowd of well-wishers around King thinned out, Collins introduced himself.

"Welcome to chromosome 17," he said with typical wit while offering his hand to a startled King. With those words, Collins and King embarked on one of the more celebrated scientific collaborations since James Watson and Francis Crick had joined forces forty years earlier to solve the structure of DNA. As with the more famous duo, the King and Collins alliance was neither simple nor staid. Until it broke up almost exactly four years later, the relationship was often warm and close, yet at the same time intensely competitive.

Beginning with a few modest but generously offered bits of advice that evening in Cincinnati, Collins aggressively pursued King over the next few months, calling her every few days. The two met a few weeks later at a private gathering of top geneticists at the Cold Spring Harbor Laboratory on New York's Long Island, the biology research institute directed by James Watson. To Mary-Claire, it seemed a chance crossing of paths. It wasn't.

"Francis was shy but he wasn't hesitant," King recalls. "Francis wanted to collaborate and I was uncertain. I soon learned that when Francis wants something he goes out and gets it."

Prior to the meeting at Cold Spring Harbor, Collins asked Watson to intercede with King on his behalf. Then 62 years old, Watson was no longer making direct scientific observation, but instead was widely respected and sought out by his peers for his advice and support. He had transformed Cold Spring Harbor from a respected but sleepy collection of research labs on the water's edge of northern Long Island into a giant in the field of molecular biology. His trademark was searing honesty, wit and intelligence, which he elegantly displayed in his acclaimed 1968 book *The Double Helix*, an autobiographical adventure describing the scientific race he won with Crick to solve the DNA structure. King's generation of human geneticists ("We are the children of Jim Watson," she once remarked) revered Watson. He returned their trust by aggressively offering ideas and opinions, often with bewitching persuasiveness.

Watson had met King only a few months before her Cincinnati presentation, in May of 1990, at a scientific conference in Spain. Actually, the two had been introduced at several genetics meetings over the years, but Watson had no recollection of her. During one set of presentations in Spain, however, the two sat side-by-side in the audience. At one point, King leaned over to Watson and commented on the speaker: "Isn't he terrific?" she said.

"Watson looked at me as if astonished that I was speaking to him at all," King recalls. "He didn't remember I existed."

But as the two left the session for an intermission they were met in the hallways by a common acquaintance, who asked King for a progress report on her research. Only a few weeks before, King's lab had come upon the first hint of the gene's presence. "I think we have just mapped a gene for familial breast cancer," she blurted out. Unknown to King, Watson was standing behind her, and upon hearing what she said he put his hand on King's shoulder, abruptly turned her face toward his and, with a look of disbelief, asked, "You did what?"

During the break, over several cups of coffee, King filled Watson in on the tribulations of her fifteen-year quest, and the breakthrough she believed had just occurred. "I remember thinking, 'Maybe she has a chance of being right,' " Watson recalls. "I did think that if it were true, then this lab I had never heard about before had just made one of the most important discoveries in gene mapping, maybe in all of medicine."

Later that same day, King presented to the meeting research unrelated to breast cancer that was the subject for which she had been invited to the seminar. For several years, King was using a DNA fingerprinting technology to help solve political crimes in Argentina. Starting in 1981, King assisted a human rights group, the Grandmothers of Plaza de Mayo, struggling to locate children who had been surreptitiously kidnapped from their families by the Argentine military junta during an eight-year period in the 1970s and early 1980s. By analyzing genetic material from relatives of the missing children, King's lab in Berkeley proved that many infants had been given to the junta's supporters after the children's parents were executed for various anti-junta offenses.

The account of her Argentine experience was a typical King tour de force, a heartrending performance that she dramatized with a sometimes-trembling voice and photographic slides of mothers who had disappeared and of their young children who had been reunited with grandparents and other relatives. Watson was among the many at the meeting who had known nothing at all about the Argentine work. As King returned to her seat next to him, Watson turned to her and said, "You know, you're quite a girl." King laughed and said, "Thanks. You're quite a guy."

From that moment on, Watson was enthusiastic about King and her breast cancer breakthrough. But he doubted that she and her small inexperienced band of graduate students had the technical skill to take the discovery further, to isolate the gene from the 100,000 or so human genes that make up the human genome. King was originally trained as a mathematician, and her advanced degree was in a field known as population genetics, a discipline that Watson did not believe prepared King for the arduous task of identifying the makeup of a hidden gene. Population geneticists study families to see if a particular trait, such as a disease trait, is being passed from one generation to another in a pattern dictated by the natural laws of inheritance. If the disease pattern conforms to the inheritance rules, then scientists can claim that the cause of the disease is a deleterious gene, even without knowing which gene it is or how it works.

"I fell into genetics because I love to solve puzzles," King says. "There is no greater puzzle in the world than piecing together the origins of human traits." It was her pattern recognition skill that first led her to believe so firmly that breast cancer was a genetic problem. "Anyone who looked closely could see that family history was a crucial predictor of breast cancer," she says. "Others, of course, saw that, too. Maybe I believed in it more. Maybe I was just naive or stubborn enough to believe it was a puzzle I could solve."

But King readily conceded that her strength was not molecular biology, the combination of complex, magiclike laboratory tools and procedures that scientists use to pick through the almost invisible world of the DNA code in search of a gene's precise physical makeup. Yet, her announcement in Cincinnati immediately set off an international race among the world's most-talented molecular biology labs.[1] The gene-hunting scientists were continually on the prowl for high-profile targets at which to aim their new laboratory weapons, and they were especially drawn to projects that by their very nature would attract the attention they needed to help fund their research. No gene had more allure than the one King now said existed. King had pointed the way, and the other labs were off and running before King's lab could even catch its breath. If she and her lab were going to compete in the race, she was going to need help.

No one understood the importance of the breast cancer gene more than Watson. In 1989, he was named the first director of the U.S. government's National Center for Human Genome Research (also known as the Human Genome Project), a federal agency set up to identify the precise sequence of all the three billion letters in the human DNA and the mapping of all the genes that are produced from those letters.[2] Watson was a master salesman and politician, and he knew that isolating a breast cancer gene would bring the kind of attention the young project needed to persuade Congress to keep federal dollars flowing. The breast cancer gene "is a very big prize," Watson said. "Something this important is certain to attract the very best in the field. And that's as it should be. Competition is very important to the advancement of science. Very often, the keener the competition, the better the science."[3]

During the meeting at Watson's Cold Spring Harbor laboratories a few weeks after the Cincinnati speech, the Nobel laureate gently drew King aside and asked her how she and Collins were getting along. Watson believed a collaboration of the King and Collins labs was a logical next step, and he encouraged Mary-Claire to accept it. In response, King opened her heart to Watson, pouring out a litany of her hesitations and fears. "I told [Watson] I was worried that in a collaboration with Collins I would be completely outclassed," she says. "I'd never screened DNA libraries, done mutation analysis, isolated RNA, run northern blots. I'd never cloned a gene. I said, 'This guy Collins is a whiz at all that.' "

But King kept her most serious concern about Collins private. "People said he was very ambitious," King recalls. "I didn't have problems with that. I did worry that in a collaboration with Collins I might become a helper, not an equal. I was adamant that my lab be an equal partner. After so many years of hard work, I couldn't take that chance."

Watson understood her reservations about Collins's superior technical talent. Still, Watson believed firmly in collaborations. And he felt it was his job to bring research scientists together, especially a dazzling duo of intellects like Mary-Claire King and Francis Collins. Their expertise in different disciplines produced a sum much greater than its parts. Watson, of course, knew the power of such alliances. He had benefited from a famous union of seemingly unrelated sciences. Watson was a biologist, but he told King that it was Francis Crick's mastery of physics that was crucial to the discovery of DNA's twisted, helical shape.

"I needed a Francis," Watson told King, "and so do you."

The poetry of Watson's argument, as much as anything else, finally swayed her. Collins had agreed that even if the gene's identity was confirmed by researchers in his lab, both teams would share equal credit for the find. Moreover, he promised to provide valuable technical assistance to King and her co-workers, teaching them some of the specific techniques and strategies Collins had helped pioneer. "By the time we agreed to work together I definitely felt the risks of collaborating with Francis were well worth taking," King says. "That doesn't mean I wasn't wary. I was."

But just a few weeks after the Cold Spring Harbor meeting and before King formally acceded to the overtures from Collins, something else happened to persuade King she had better join up with the tall, ambitious biologist if she were to retain her lead in the nascent field of breast cancer gene research. By the time King agreed to join forces with Collins, a host of colleagues in other labs were breathing down their necks.

Unknown to Mary-Claire King, across the Atlantic Ocean in France, one of molecular biology's foremost practitioners already was prepared to test the validity of her claim. Within days of her announcement, King received a telephone call from Gilbert Lenoir. "I was pleased, but also anxious," King says of what Lenoir told her.

Lenoir is a professor of genetics at the University of Lyons, directs France's International Agency for Research on Cancer, and is one of France's most talented gene-hunters. In 1985, his lab succeeded in identifying a mutated gene at the heart of an inherited cancer called endocrine neoplasia. Although the disease produced tragic illness in people at a young age, it was rare—only about 150 families suffered from it in France. "We decided if we were going to have an impact we should use the same [gene-hunting] techniques to go after really big cancers," Lenoir says.

Back then, Lenoir knew the key to success was amassing the right collection of families. Once his lab was armed with family DNA, his lab knew how to scan it for a disease-causing gene. He began his search in a large family from Iceland, but got nowhere. Then he heard about Henry Lynch, an obscure doctor in Nebraska who had been studying familial cancers since the early 1960s. A cancer specialist, Lynch believed that the disease sometimes involved inheriting a damaged gene.

The evidence, Lynch argued passionately, was there if only doctors paid attention to their patients. "You take a family history and often you find all these other people have the disease, too," Lynch argued. "If it were any other disease than cancer, the doctor would say it looked to be inherited. The notion that cancer could be passed from parent to child was just not an acceptable scientific idea." But over the years, Lynch and his colleagues tracked down dozens of families in the United States with breast cancer and dozens more with colon cancer. Lynch even christened the previously unseen familial colon cancer after himself, dubbing it the "Lynch syndrome." Despite having published scads of research articles, Lynch found that his colleagues in cancer research brushed aside his claims. Few geneticists even were aware of his work, and those that were paid little attention to it.

But when Lenoir decided to go after a breast cancer gene in 1986, a colleague suggested he contact Lynch at his Creighton University School of Medicine office in Omaha. For days, Lenoir examined the descriptions of the women in dozens of families listed in a thick registry Lynch had compiled over more than two decades. "I was amazed at the data," Lenoir says. "It was as if I'd stumbled into a gold mine." In family after family, breast cancer arose on average in about half of the offspring of a woman with the disease. "It had all the characteristics of a dominant gene," Lenoir says. "I never realized such data existed."

Lenoir was particularly fascinated by a subset of five families that shared three distinctive characteristics he guessed might be caused by the mutation of one potentially identifiable gene. Half the women in these families developed breast cancer before age 40, as compared with about 5 percent of the women in the general population. Many of these women also developed cancer in both breasts. "Bilateral disease," meaning cancer of both breasts, suggested an especially pernicious form of the cancer that might result from a shared inborn defect, as opposed to the battering over time from an environmental agent, he guessed. And, most striking, the families had an unusually high rate of ovarian cancer.

"I knew [breast cancer] was complicated, that it was probably a result of many factors, perhaps many genes," says Lenoir. "I was looking for a way to simplify the search. These families' shared characteristics seemed to hold the key." By 1988, Lenoir had enough genetic samples from the families in his lab to embark on a search for the gene.

By the autumn of 1990 when King surprised the world genetics community, Lenoir's group in Lyons had spent almost two years unsuccessfully marching up and down many of the 23 chromosomes. When King reported a gene on chromosome 17 "we already had everything in place to test her claim," Lenoir says. Indeed, Lenoir was in Cincinnati, attended the late-night session, heard King's talk, and the next day called his colleagues in Lyons, directing them to begin probing for the gene in the Lynch families exactly where Mary-Claire had shined a light, on chromosome 17.

Every week or so for the next few weeks, King received an overseas call from Lenoir asking for advice or seeking some bit of technical information. "Each time I asked him, 'How's it coming? How's it coming?'" King recalls. "He kept saying, 'I don't know.' The longer it took, the more concerned I became that maybe we were wrong. It was a very tense time."

In early January, two weeks after the journal *Science* carried the report on chromosome 17, King was invited to London to join a group of esteemed gene hunters to discuss strategies to confirm the Berkeley claim and close in on the gene if it existed. Among those invited to give a presentation was Lenoir. King asked him what he was going to say, but Lenoir put her off. "Gilbert was very dour, very quiet," King says. "I thought, 'Oh rats! He's found something that's going to blow us out of the water.' Gilbert is a great actor, very dramatic. Before he began speaking he asked me privately what I thought the chances were that I was right about chromosome 17, and I told him, and I meant it, that I thought maybe I had a fifty-fifty chance. He looked at me with a very straight face and said, 'Maybe.' I didn't know what to make of it. Really, the guy was driving me crazy."

King sat down in the small meeting room, the lights were turned low and she braced herself for bad news as Lenoir began discussing several families, illustrating each pedigree with a slide. To King the slides looked familiar, as did statistical data Lenoir also presented. King was confused, and assumed Lenoir was first describing her lab's discovery as a way of contrasting it with his data. "So, Mary-Claire," Lenoir asked, "how do you like these results?"

King said, "I like them fine. They're our results, aren't they?"

"No," said Lenoir. "They're my results."

King let out a shriek. "Are you kidding?" King asked. "Are you finding what we found?"

Once more Lenoir said, "No." He then added dryly, "Our results are slightly better than yours." While King's lab had suggested the gene caused an inherited form of breast cancer, Lenoir said the same gene also placed carriers at an abnormally high risk of developing ovarian cancer. In a formal report of his finding published months later, Lenoir wrote: "We have been able to strengthen the initial claims for linkage for hereditary breast cancer with an independent set of families."[4]

After Lenoir finished speaking, Ellen Solomon, the meeting's organizer, suggested the scientists break for tea. Solomon was a respected biologist at the International Cancer Research Fund in London, where the meeting was being held. Teaming up with her mentor at the center, Sir Walter Bodmer, Solomon had made numerous key insights linking inherited genes and cancer. "Lenoir's talk was very convincing," Solomon says. "There was still a lot of lingering doubt before that. Gilbert's work was very impressive. But, of course, Mary-Claire had gotten there first. It was a great, great moment for her. Over tea we all congratulated them both. None of us said it but we all knew that everyone in that room was going to go home and begin to chase after Mary-Claire's gene."

King realized that, too. But for the moment she simply felt relief. Later, on the plane home and when she reached the comfort of her lab and colleagues and friends, King allowed herself to put her feet up and savor her triumph. "In those days, before the competition heated up, I was able to think about what we'd accomplished," King says. "It's a popular fiction that I knew all along what I was doing. It really was not a linear path at all. In fact, there was a great deal of stumbling, of false starts. You know, when I was young and just starting out, there really were very few women in the field. We weren't taken very seriously. So when I declared that I was going to find this gene I don't think I was taking that great a risk. Nobody expected much of us; I didn't expect much either. I had the luxury of failing, and of failing over and over. Not until Gilbert confirmed my finding did I finally feel I'd made a contribution. And then, you know, I felt a responsibility to do more."

4: Family 15

Even before the gene itself was identified, Mary-Claire King's discovery was unleashing a power no one was prepared to handle, including the two Michigan sisters, Janet and Susan.

As a direct result of the collaboration between King and Francis Collins, in late August of 1991, the two sisters found themselves sitting in a small doctor's office at the University of Michigan waiting to hear news of their future. Did Susan have the gene or not? Was she destined to fall victim to the cancer that had already struck her mother, aunt, sister, and cousins? Did surgery to remove her breasts make sense or not?

To Barbara Weber, the moment seemed excruciatingly long. Susan had said she wanted to know what data the university researchers had in their hands, information that would surely change the course of her life. Finally, as Janet and Susan sat close to one another, holding their collective breaths, Weber said: "We don't think you should have the surgery. It's not necessary. It's absolutely not necessary," Weber said, smiling broadly. "You don't have the gene."

Susan and Janet were in each other's arms before they knew it. Bill started to cry, and tears ran down Weber's cheeks. The next few moments, and much of the next few days, was a blur to Susan. "I had trouble hearing what people said," she remembers. "I had trouble talking. All I remember thinking is this can't be true, and wondering if I really should trust what they were saying."

Susan and her family were being confronted by two facts unknown several weeks before. For one, they were being told that the cancer that had stalked their family plaguelike was the result of something they had inherited. Susan now realized she had known this all along, although she couldn't name it or describe it scientifically. Now there was nothing to face but the truth. Yet, at the same time, she was being told that unlike her beloved sisters and cousins, she was freed from a fate she had believed was inevitable. For the past few years Susan had organized her life around the idea that she would soon be fighting cancer, and that the chances she would win that struggle were slim. Instead, she was being told, she stood the same risk as any other woman.

Why was I spared? she wondered. "Why did Lucy and Janet get this gene and I didn't?" she asked Bill and Janet on the ride home that evening. It doesn't make sense. How is something like this decided? What did I do right? What did they do wrong?" Instead of feeling exhilarated, she was uneasy. Only later did she understand how guilty the news made her feel. "How could I enjoy what they had told me?" she says. "I was happy for everyone else in the family because they wouldn't have to worry about me. And I was relieved, very relieved that my [10-year-old] daughter would never have to go through the worry I'd lived through and that she would grow up healthy."

Indeed, the only exhilaration she felt came whenever she thought about her daughter, Jessica, then 10 years old. Only a few weeks before, Jessica had made a disturbing announcement. As she prepared to take a shower, she told her mother, "I don't want to develop [breasts]. I don't want them." Susan couldn't wait to get home and tell her daughter she needn't worry anymore about growing up. Because Susan didn't carry the gene she couldn't pass it along to her daughter. But "to this day I can't shake the other feeling," Susan says. "Why did God spare me?"

The news of what occurred in Weber's office sped through the cancer clinic. Virginia Martin, Weber's nurse, remembers feeling "awestruck." Says Martin, "At first, I didn't believe it. I wasn't that familiar with the technology and I thought, 'Gee, what if we're wrong?' For a moment I was a bit light-headed and had this tingling feeling all over. I think everyone in the clinic who heard what happened was a bit overwhelmed. None of us knew this kind of information was being uncovered."

In the next few days the Michigan doctors and scientists realized they had created a dilemma whose true dimensions didn't become clear until months later. Even several years afterward, Weber and her research colleagues were still debating how well they handled the events in the weeks and months that followed. "We'd do things now a great deal differently," says Barbara Bowles Biesecker, a genetics counselor recruited to the team a few days after the gene link was revealed to Susan. "But this was the first family we had to deal with. Now we know there are other ways to inform the many other families that will have to face the problems we first ran into that summer and fall. We think we learned a lot, but I think we have a great deal more to learn."

Says Kathy Calzone, another nurse-counselor on the Michigan team, "We have yet to provide the family with the emotional support they need. None of us were prepared for the problems that came up. I don't think if we sat around all day we could have predicted some of the things that occurred. The truth is [the family's] needs have exceeded our capacity to deal with them."

Janet and Susan went home and shared the news with their mother and two brothers. Within a few days, Susan's tale spread quickly to a widening circle of relatives. "Right away, we realized we had a problem," says Barb Weber. "Susan and Janet called and said other people in the family wanted to know where they stood regarding the gene. From what the two sisters were telling us it was clear there was a great deal of misinformation being passed around. We decided we had better tell those people who wanted to know if they had the gene."

"But do you know what that meant?" Weber asks. "We were going to tell some young women that some time in the not very far-off future they almost definitely were going to have breast cancer.

Just like with Susan and Janet, some sisters would be spared and others wouldn't. We wondered. 'How old does someone have to be to hear this kind of news?' And then what? There was no treatment on the horizon for someone with the gene. We could only suggest close monitoring to catch the disease early, nothing that could stop it from occurring. What kind of help was that? Should we recommend preventive surgeries to healthy women? And at what age? This definitely is not what we had intended when we started the research."

Weber and her boss Collins had stumbled into a new "gene-age" experience that, they now saw, would soon be repeated in other diseases for thousands, perhaps even millions, of families around the world.

Susan and Janet's extended family first entered this new "gene-age" in early 1991 in a clue hidden in a cancer registry. Kathy Calzone, the nurse recruited to assist the university scientists, was rummaging through the medical center's data base, scanning for breast cancer patients who also had one or more close relatives with the disease. In June, after months of looking, Calzone came across the medical records of a potential research subject. The records belonged to Dolly, the mother of Janet and Susan.

Calzones's discovery was far from trivial. The key to Mary-Claire King's discovery was her collection of about two dozen families where the cancer struck three or more "first-degree" relatives: sisters, daughters, mothers, grandmothers, or aunts. It took King years to mass her impressive collection of families with this kind of profile. As a result of her family compilation, scientists now believe about 5% or so of all breast cancers are inherited.

"Mary-Claire's contribution is almost impossible to measure," says Francis Collins. "She proved the gene existed and she pointed the way to where it lay. But there was still a big job ahead. If you consider human DNA as being the size of Earth, King had just placed the gene somewhere in Texas. It was no small accomplishment because few people even believed it had existed at all. Yet, now the job ahead was to pinpoint the gene, to first map it to a particular county in the state, then to a town, then a street, then a house on the street, and finally, the exact room in that house. It was going to be a very difficult assignment and that's why it was so important that so many top molecular biology labs had joined the effort."

In order for his laboratories to join in the hunt for the gene's exact identity, Collins then needed DNA from families beset by a heavy incidence of breast cancer. Sharing DNA from such families was something researchers in human genetics did reluctantly. The families were invaluable resources, too hard to come by and too rare and too precious to be passed around from lab to lab without guarantees of something valuable in return. Collins had access to some family DNA to begin his studies through collaborations he quickly set up in early 1991. But he realized he needed to find his own families if he was to play an important role in the race to bag one of the most sought-after and prestigious of all gene-hunting game. Barbara Weber, then 35 years old, was recruited to help. She had just finished her postgraduate work in genetics at the university and had been given her own molecular genetics lab in Collins's department. She and Collins then enlisted Kathy Calzone to find research families.

Dolly Burton's name popped out of the data base because she had been treated at the hospital and had stated that her older sister, Matilda, also had breast cancer. Nothing could be certain about Dolly, about her health, her current residence, or her willingness to participate in the research. Calzone entered Dolly's name in her own computer and, in an effort to protect the family's privacy in future discussions, identified her as the "proband," or initial contact, of "Family 15," for Dolly was the fifteenth woman uncovered in the data base whose family history warranted a deeper investigation.

Much later, Family 15 would become renowned in medical research articles and in discussions among geneticists. But that day, Dolly was just another research lead. Calzone sent her a form letter asking if she was willing to answer some questions over the phone. A few days later, Calzone received a call. Dolly had a story to tell. Indeed, Dolly was eager to unburden herself.

In November 1978, a gynecologist found a small round nugget of tissue in Dolly's left breast during a regularly scheduled health exam. It was cancerous. Dolly was sent to a local hospital where a surgeon performed a mastectomy and where she had a subsequent round of chemotherapy. The surgeons found the tumor hadn't traveled to the other breast or beyond that to her lymph nodes (small packets of hormones and blood cells located under the armpits). Doctors check there first to determine if a cancer is confined to the breast or has possibly metastasized dangerously to other tissues and organs. Sometime later, Dolly went to the university with one of her daughters for a checkup, and her name and a brief notation of her sibling's cancer were entered into the university's cancer registry, to be uncovered later by Calzone.

Dolly was 48, and, as far as she could recall, no one in her family had had breast cancer. Her father, Peter Stone, had died of cancer of the prostate gland when he already was an old man of 82. Stone was a dairy farmer, like his father and grandfather. As was the custom among many midwest families who ran their farms with the brawn of their many children, Stone was one of ten siblings. Cancer wasn't common among his brothers and sisters. But farm life was plenty hard in rural Michigan in the first half of the twentieth century and several of his siblings died before reaching adulthood. Two of his sisters were killed in a freak accident, run over by a train while riding in the back of a horse-drawn wagon whose driver lost control of his horses. Of Dolly's six siblings, one brother did die, at age 28, of what was described to family members as liver cancer. But his illness was considered an aberration.

Dolly's cancer was a shock to the family, which had moved into the prosperous middle class of central Michigan where tourism, real estate, and light industry were taking over from farming as the dominant area employers. Dolly's doctors warned her in passing that the disease was known to cluster in families, and she passed along his concern to her sisters and nieces. As a result, the following February one of Dolly's two sisters, Matilda, known in the family as Mattie, went for a mammogram, a diagnostic imaging test that can detect the presence of a knot of suspicious tissue within a breast. A lump of thicker-than-normal tissue within the breast appears on a mammogram's film as a dark bit of shadow as compared with the normal light gray of healthy breast tissue.

Mattie, a sweet, stout-looking woman, good-naturedly referred to her breasts as being "massive." Her mammogram displayed an unmistakable shadow in her right breast. But because of her breast size, neither she nor her doctor was able to feel a lump in the area identified by the test. "But there it was growing inside of me," she says many years later. "If Dolly's doctor hadn't suggested we all get mammograms I'm not sure I would have caught my cancer so early." Surgeons removed the entire breast. As with Dolly, there was no evidence the cancer had spread.

Mattie was 50 years old. The fact that two sisters had been struck by cancer at about the same age, and within several months of one another, sent a shiver through the family. This double blow, the women believed, was no coincidence. Mattie and Dolly implored their daughters to get examined, too, although the oldest, Pamela, was only 30 years old, far younger than normal to get breast cancer. That's why when Pamela finally got around to taking a mammogram six months later, the diagnosis of cancer came as a shock.

But Pamela's cancer also had not spread. Her breast was removed and, after a time, all seemed to go well with the family. In the early 1980s, "we took what had happened as just odd, you know," remembers Janet. "We didn't talk about it much. We were all having children, we visited with each a lot and the whole family got together on Christmas. We made cookies together, exchanged gifts; our husbands hung out together, hunting and fishing. Our lives seemed very good."

Then in 1985, Pamela was diagnosed with ovarian cancer. It was picked up in a normal gynecology examination that included a Pap smear, a test in which a doctor scrapes cells off the inner walls of the uterus and sends them off to a clinical laboratory where they are inspected under a microscope. Many of the cells in Pamela's test were abnormal, having changed into shapes and sizes typical of cells that were cancerous.

It was the begining of a harrowing four years that changed the family forever.

Ovarian cancer is more deadly and difficult to treat than breast cancer. It strikes about 30,000 women each year in the United States, far fewer than breast cancer, but the rate of survival among women after five years is much lower. What frightened the Stone women about Pamela's newest cancer was that doctors told them it was unrelated to her breast cancer. It was not a result of malignant cells' migrating from the breast.

"When we heard about it, we all thought [her breast cancer] had spread to her ovaries," Janet says. "When her doctors said it was totally different, I thought, 'What? Another cancer? Not part of her breast cancer.' All of us started thinking, 'This is so strange. I mean, what the hell is going on?' "

Stranger still, within the year, another of Mattie's three daughters, 29-year-old Beth, who lived in the South, was also diagnosed with breast cancer. And the news right from the beginning was bad. Within a few years, the cancer had broken loose from the breast, drifted into the her bloodstream and then lodged itself in several sites elsewhere in her body. Soon Beth's body was freckled with tumors. Beth, her husband, and two children moved back to Michigan where she began her own desperate fight against the disease and its swiftly ravaging attack.

By this time Pamela's cancer had spread beyond her ovaries. Doctors at a nearby hospital administered doses of chemotherapy and radiation so intense that within a few months Pamela fell seriously ill. The anticancer drugs damaged her bone marrow, cells deep within bone tissue whose job it is to produce and churn out the body's supply of red and white blood cells. With her marrow dangerously depleted, Pamela became anemic and susceptible to life-threatening infections. In an effort to replenish her devastated marrow, Pamela was entered into an experiment at M. D. Anderson Cancer Center in Houston, where doctors were testing a new family of drugs that stimulate the body's manufacture of red and white blood cells. Leaving her husband and two teen-aged children at home, Pamela and her mother and father traveled to

Houston for treatments that ran for three weeks every month, for five months.

While at M. D. Anderson, Pamela came to believe the cancer outbreak "was inherited," says Mattie. "I heard her argue with her doctor about it once. She said she knew it was something in the family. And she begged the doctor to help them find out what it was. But the doctor said they had no way to know for sure. He said it could be anything."

Believing the women were under siege from some unseen but unstoppable force whose genesis was inside them, Pamela implored her other sister, Jesse, to take the offensive before it was too late.

"Pamela kept after her," Mattie says. "She told her over and over, 'Remove your breast tissue, remove your breast tissue.' " But, quite understandably, she resisted. The idea of excising healthy tissue just didn't seem rational. As the youngest of Mattie's three girls, Jesse didn't take kindly to being bullied. She decided to hold off.

But then in late 1986, the third case of cancer in less than two years descended on the family. This time it was Jesse's cousin Lucy, the oldest of Dolly's three daughters and the elder sister of Janet and Susan. With two sisters dying of the disease, and a first cousin diagnosed with breast cancer, 27-year-old Jesse finally capitulated. She ordered surgeons to take away her healthy breasts. It now seemed rational to remove her breasts and crazy to keep them.

Lucy's tumor first surfaced as a tiny mass of hardened tissue she discovered during a self-examination of her left breast. Lucy had a common condition called cystic fibroid disease, in which harmless lumps of tissue often form throughout the breast. The problem, however benign, worried her doctors because it made it difficult to detect whether a tumor was growing alongside the cysts. As a precaution, therefore, Lucy's doctors recommended that she have her right breast removed, along with her cancerous left one. When tissue from the right breast was examined under a microscope, the doctors weren't surprised that the cancer already had spread. That breast was also removed and Lucy underwent breast reconstruction.

One day while Lucy was planning to undergo final stages of plastic surgery, she noticed a small bump under the skin atop her left-side collarbone. The cancer had metastasized. By Christmas, doctors found it already had spread to her lungs. And a few weeks later, tests found it also had invaded her brain.

"Every night I cried myself to sleep," Susan, the youngest of Dolly's three daughters, recalls. "Lucy was like a second mother to me. She understood me like no one else in the family. We talked on the phone every day. She was beautiful and smart; she knew the right things to say to make people feel good. She was the glue that held the family together, and we all depended on. I hated to see her suffer. I don't think I ever was so angry about anything."

One day Susan visited Lucy and discovered her sitting intently on her living room floor with about a half-dozen large x-ray films spread around her. Lucy was looking at her brain scans, x-rays of her brain where new evidence of the spreading cancer showed up as bright spheres, small but unmistakable colonies of massing cells deep within the most sensitive of human organs. Their location was simply beyond the reach of the surgeon's scalpel. "I just wanted to hug her and make her feel okay," Susan says. "It was then I realized how lonely she must feel. There was nothing I could do to make her pain go away. I couldn't even begin to imagine how she went through her day, how she fell asleep at night. I couldn't bear to think of her pain."

Cancer was now an everyday topic; it could no longer be avoided or denied. The women traded advice about chemotherapy, radiation, and biopsies. They became conversant in the stages of the disease, in experimental protocols and where they were being offered. They learned about the varied types and dosing of anticancer medicines, and discussed the relative side effects of cisplatin, methotrexate, and tamoxifen.

In February 1988, Pamela's battle with the disease ended. At Pamela's funeral, Susan couldn't take her eyes off her sister Lucy. By then, Lucy was wearing a wig; her hair had fallen out under the onslaught of a new round of chemotherapy. "Don't give up, Lucy," Susan whispered to herself. "Don't give up."

Ten months later, just two days before Christmas, Mattie's other stricken daughter, Beth, slipped away. Three months later, Lucy died, too, leaving behind her husband and two teenage daughters. The Stone family women were now in shock, alternately angry and deeply depressed. Janet and Susan were terrified beyond words at the wreckage around them. Janet found herself wishing to turn back the clock, and choking back spasms of unexpected rage at the least little annoyance. "All I could think about were the days when we were younger, when we started having children and we'd sit around each other's homes or we'd babysit for each other," Janet says. "Those days were so carefree. It just never occurred to us that we wouldn't spend our entire lives together, watching our children grow up. Who would ever have thought we would be robbed this way? When it started to happen, we wondered, 'What did we do? We're Catholics and I stopped going to church. I couldn't go. I was thinking, 'Screw you, God, screw you!' "

5: Trick or Treat

When Kathy Calzone, the University of Michigan researcher, contacted Dolly in early 1991, Susan and Janet were expecting that any day they would become victims of the disease. The women debated whether they should have "prophylactic" double mastectomies. Susan's doctor also was worried. Whenever he saw Susan he urged her to remove her breasts.

After Calzone spoke with Dolly and then Mattie, she believed the family's DNA was worth mining in the search for a gene. In order for the scientists to begin sifting through the family's genetic material, Calzone was directed by the lab scientists to retrieve blood samples from each woman affected by the disease, from their siblings, and from both parents. Not surprisingly, this wasn't going to be easy. Mattie, Dolly, and their unaffected sister, Mary, agreed to give their blood, as did their 91-year-old mother. But blood from their father was unavailable; he had died ten years earlier. Janet and Susan agreed to give blood as they were the sisters of an affected individual, Lucy. But Lucy was also deceased by now. The same was true of Mattie's two affected daughters, Pamela and Beth, but their surviving sister, Jesse, agreed to participate.

The Michigan researchers decided to move ahead despite the hurdles posed by the missing samples of the deceased women. As in the experiment involving the mystery of Hubert Humphrey's cancer, the researchers held out hope that they eventually could track down stored tumor tissue from the dead women, from which DNA might be extracted.

In October, Calzone mailed the family a large sampling kit with six test tubes, prepaid return mailers, and a lengthy set of instructions. It told them to go to a personal physician or to a nearby medical-testing laboratory to have the blood drawn. Once filled with the women's blood, the test tubes were placed in sealed plastic bags, then packaged into special Styrofoam-encased overnight mail pouches addressed to Calzone.

In order to get the family's cooperation, Calzone explained that by studying DNA from families similar to theirs, the genetic basis of their disease might someday be revealed. "I didn't give it too much thought," says Janet. "We were glad to be doing something to stop the disease, for us and others. But I wasn't expecting it to help us any."

The researchers then asked the family to undertake some grisly detective work; they had to find whether any remnants of the dead relatives still existed. The family contacted their doctors and provided Calzone with a list of medical centers where their relatives had been treated over the years. And during the winter, Calzone first tracked down the hospital where Susan and Janet's grandfather had his prostate cancer removed, and then, one by one, Calzone identified the medical centers where tumors were extracted from Pamela, Beth, and Lucy.

"It took months and months to get people at the hospitals to figure out if they had [stored] tumor blocks from the deceased relatives," says Calzone. "We were lucky. Tissue samples from all four of the relatives did exist and they were available. But then it still took time to get the necessary consent forms and other agreements in place. We didn't get all the samples together until the early summer [of 1992]."

Meanwhile, Janet and Susan were vigilant. In November of 1991, the women traveled together to the Mayo Clinic. "We decided to have some of our [six-month] checkups there," Janet says. "Maybe it would change our luck." The good news continued, although the sisters both felt it was simply a matter of time before a cancer turned up.

Then, during the following Memorial Day weekend while visiting her inlaws in Cleveland, Janet found what she had been looking for. She had gone to bed alone while her husband, Tom, stayed up to have a late-night talk with his father. "After my period passed I always did a self-exam," Janet says. "I did it religiously even though the idea of it made me nauseous." In the darkened guest bedroom while lying quiety in bed, Janet first ran her right hand over her left breast. Then, while moving her left hand across her right breast just under her armpit, she stopped. "Oh, my God," she thought, not really believing what she felt. "There's a lump there!" For a moment she lay still, stunned and not thinking, quickly touching the same spot over and over to be certain she was right. "Well," she thought, "I don't have to worry about this anymore."

Her thoughts confused her, but she felt unexpectedly calm. "I knew I had cancer," Janet says. "But I didn't feel upset. I felt resigned." A few minutes later, Tom came into the room to kiss Janet good-night and she asked him to touch the same area. He did, and then looked at her in darkness and said, "I don't feel anything." Later he would tell Janet that he did "feel something" but that he didn't want her worried. Instead, he quickly returned to his father. But he was concerned.

Remarkably, over the next few days Janet and Tom hardly discussed the lump. The following weekend Janet had organized an elaborate party at her home to celebrate her daughter's high school graduation. "I don't have time to deal with this now," Janet thought of the lump. Instead, she decided to wait two weeks for a scheduled appointment with her gynecologist to conduct her six-month breast examination. "I decided not to do anything, I guess not even to think about it, until I went to see the doctor," she says.

Janet's physician was casual and reassuring during the exam, telling her that to his experienced touch the ball of tissue under her right armpit didn't seem rigid, pointed, or hard, the kind of feel he associated with a tumor. Instead, it was soft and movable.

"If it were anyone else," the doctor told Janet, "I wouldn't be worried at all. But with your family history I think we ought to do a needle biopsy." Janet wasn't surprised that he recommended doing the biopsy right there in his office. It was Tuesday, and the doctor told her he would call her with results the next Monday.

At home that evening she "put up a big front for Tom," she says. "He was upset that I had the biopsy done at the doctor's office. He felt this kind of thing should be done by a specialist. I told him I didn't want to wait." Janet went to work on the next day. But then on Thursday, while at work, she became distraught. "I knew I had cancer and I knew I had to find out the truth," she says.

The next morning she stayed home from work and called her doctor's office. Each time she called she was told that his nurse was on another line. The nurse was calling Janet, and each time she tried, Janet's line was busy. When they finally connected in the late morning, the nurse asked Janet to come to the office right away and bring her husband. "I knew right then it was bad news," says Janet. "They'd tell me over the phone if it were nothing."

Once in his office, Janet's doctor didn't mince words. "I guess I don't have to tell you why I have you here," he said. "I think you already know."

"Yeah," Janet said. "I know."

"It looks to be very early, and I think your prognosis is very good," the doctor said. "But you need to find a surgeon and you need to have surgery."

Because breast cancer was no stranger to the family, few who heard about Janet over the next few days needed to ask what it meant. Just like Janet, they knew. When Janet's daughter Meg came home for lunch that day, she sensed something was wrong. Janet's cheeks were puffy; her eyes were red from crying. Without dragging out a long explanation Janet said, "I have a lump and it's malignant." Meg, a radiant young woman, instantly burst into tears.

Her daughter's reaction yanked Janet up from the well of self-pity into which she so quickly had fallen. Instantly, she saw that her cancer was likely to devastate a family whose emotional girding was already weakened. "When I looked at Meg's reaction I thought how different it was from the way kids in most families would have acted," Janet says. "She didn't ask me what it meant. She didn't assure me, or look for me to tell her I'd be okay. Instead, she started crying hysterically. She couldn't stop crying and there wasn't anything I could say to help her. I wanted to tell her I was going to be fine, but I didn't know if I was. We'd all watched my cousins and sister die. We knew what this was. Later, my [13-year-old] son came home and when I told him he reacted the same way; he started to cry. It was so heart-wrenching. I felt like I let them down."

A few days later, Janet met Barbara Weber at the University of Michigan medical center's oncology clinic. Janet's gynecologist had advised her to seek out a surgeon at the university's hospital. As a first step she needed to be seen by a staff cancer doctor, or oncologist, to review her medical chart and lay out treatment options. The oncologists at the hospital often were assigned new patients on the basis of which doctor was available when a patient arrived for her first consultation. Janet was assigned to Weber, who was staffing the clinic that day, and the two faced each other for the first time in one of the clinic's small examining rooms.

Research studies in recent years have shown that small cancerous lumps the size of the one found growing inside Janet's breast can be successfully treated by snipping out a small bit of breast tissue surrounding the cancerous mass. Studies show this procedure, called a lumpectomy, works as well in small tumors as does a radical mastectomy, in which the entire breast and surrounding muscle tissue are removed. Weber explained all this to Janet, but quickly realized she was dealing with an unusual patient. "Her sophisticated knowledge of treatment options, and what she wanted to have done to her surprised me for a moment," Weber says. Janet didn't want a lumpectomy or a mastectomy. She told Weber she wanted a "bilateral mastectomy," that is, she wanted both breasts surgically removed.

It took Janet only a few moments to chronicle the family's thirteen-year saga. The family history sounded familiar to Weber, and it was then she realized that fate had brought her together with a member of "Family 15," one of the handful of research pedigrees whose DNA was being collected by Weber's colleague Kathy Calzone.

"Janet and her family were now my concern, not as a removed researcher, but as an intmately knowledgeable personal physician," Weber recalls. "The need to help them in some way, any way, became foremost."

Soon afterward, Janet had the surgery. Doctors removed both breasts even though there was no sign the cancer had spread beyond the small, dime-sized tumor she had discovered a few weeks earlier. Janet asked for aggressive treatment and was assigned a hefty regimen of cancer drugs to be administered to her as an outpatient twice a month at the university hospital. It was the day after receiving her second monthly dose of chemotherapy that Janet learned obliquely over the phone from Weber that scientists at the university had uncovered a critical clue about the family's swarm of cancer cases.

About the time Janet first met Weber, laboratory scientists working with Weber and Francis Collins, Weber's boss, had finally gathered enough DNA from living and deceased members of the Stone family to begin their quest for the gene. Almost two years earlier, the California researchers led by Mary-Claire King had identified the small region located in the 17th chromosome that was the likely home of the gene.

The gene marker that King found was identified as a specific segment of DNA that varies in length from one person to another. While minute or subtle differences in the DNA can be disastrous to a person's health or produce the kind of genetic variety that distinguishes a gifted musician from someone who is tone-deaf, scientists have found that most variations of letters in these DNA markers rarely produce any significant effect on inherited traits.

For Weber and Collins, and also for researchers in about a dozen labs in the United States, Canada, Europe, and Japan, the markers were rare beacons of light in the fog of DNA. The scientists believed that studying the markers in large families with cancer clusters would lead them through the darkness to the gene's exact location and identity, the first step in figuring out how the gene carries out its deadly deed.

By late July, the University of Michigan lab scientists reached an important conclusion about the Stone family DNA. Everyone in the family affected by the disease carried a particular-sized marker in the area where the California researchers had mapped the approximate location of the breast cancer gene. But other members of the family had inherited a different set of markers in that region—and these family members were unaffected by the disease.

To the scientists, the pattern meant that the Stone family disease was very likely caused by a gene hiding somewhere in chromosome 17 that was inherited along with the marker. The finding meant that the marker and the gene were located very close to one another on the chromosome. Every cell in a human carries two copies of all 23 chromosomes. Therefore, each cell actually has two copies of every human gene. Many proteins can be manufactured with only one working copy of a gene.

In the case of the breast cancer gene, just one functioning copy seemed to be sufficient to forestall cancer. In other words, if a person inherits one defective copy of the gene from one parent, every cell in the body contains one mutant copy of the gene. Scientists believe that if the other "working" copy becomes damaged in just one breast tissue cell, perhaps as a result of a random error during cell replication, then the protein normally produced by a functional gene is also defective, and will no longer protect the cell from becoming cancerous, duplicating over time into a dangerous tumor.

Here's how a DNA marker known to reside near a defective version of a gene can be used to test for the faulty gene's presence. Scientists know that a parent passes only one of his or her two copies of chromosome 17. If one of a parent's two chromosomes carries the defective gene, then each offspring has a 50 percent chance of getting the chromosome containing the aberrant gene, and a 50 percent chance of getting the one containing the working version. Which copy of the chromosome is passed along is just a matter of luck. Sometimes it is the damaged copy, sometimes the working copy. The markers found in the DNA allow scientists to tell whether a parent has passed along a chromosome copy with a mutated gene or the chromosome copy with a normal working form of the gene. This analysis can be done even though the actual gene has yet to be identified.

To make certain they were right, the scientists carried out one more experiment. They blanked out the names on the DNA samples from each family member and challenged lab researchers to use the markers to pick out which genetic material came from a person with cancer and which sample belonged to someone free of disease. The "blinded" researchers were 100 percent accurate in matching DNA containing the culprit gene to people with disease. The test was powerful proof that the Stone family cancer was caused by a gene whose exact identity was still unknown but whose existence was beyond doubt and, in fact, could be traced as it passed through a family. The test also meant that since the family's DNA contained the gene, the Stone family DNA was an appropriate place to target their high-tech gene-searching laboratory tools.

And, finally, the test meant that the same techniques used to prove the family was linked to the gene could be used to identify, with nearly absolute precision, who in the family, from the very youngest child to Dolly's still living mother, carried the disease gene. As Barb Weber inadvertently let slip during her conversation with Janet, the researchers could use the marker to predict who in the Stone family was likely to get cancer from inheriting one copy of the gene.

By Labor Day weekend 1992, about two weeks after Susan was saved at the last moment through the miracles of new-age genetics, the Michigan researchers hotly debated what they should do next. Janet, of course, wanted to know if her daughter Meg had lost the genetic roll of dice and had inherited the chromosome that contained the dreaded gene.

Within a few days, Janet and Susan called with some disturbing news. The two had canvassed many of their close relatives. The verdict was consistent. Many women in the family wanted the scientists to gaze into their genetic crystal balls for them, too. If there was a chance of good news, they wanted desperately to hear it. Lucy's widowed husband wanted to know about his two daughters. Mattie's three daughters had five daughters among them; some of them wanted to know, too. Moreover, the researchers were worried that some women who weren't gene carriers might be considering surgery, just as Susan had.

Also, "the family was using some very crude methods for doing their own gene predicting," says Weber. For instance, the family concluded that since the daughters of Susan's deceased sister, Lucy, both looked like their mother, they probably had inherited her defective gene. Their reasoning: Janet and her sister Lucy looked alike and they both had cancer, while Susan had very different facial characteristics.

Finally, says Calzone, "We decided once the family realized what we knew, we had a moral and ethical obligation to provide them information if they wanted it. But we didn't have a framework how to proceed. There was no model."

Slowly, the Michigan researchers understood they were taking a momentous step into largely uncharted ethical waters. The advanced science of their research labs had wrenched them, however reluctant they were, into practicing a type of "predictive" medicine few knew anything about. For several years, academics and ethicists argued over thorny issues they expected to emerge from the gene revolution exploding around them. Several years earlier, scientists strained the known boundaries of medicine, genetics, and medical ethics when they located a marker for the gene that causes Huntington's disease, a rare lethal disorder with a bizarre twist. People who carry the Huntington's gene live normal healthy lives into their early middle age. Then their health quickly deteriorates as the defective gene mysteriously and monstrously wreaks havoc, stealing the nimbleness of its victim's mind and eating away at the sinews of the body.

Almost immediately after the gene marker for Huntington's was discovered in 1983 controversy raged over the appropriateness of testing young adults in Huntington's families. As with Susan's experience, tested people who learned they lacked the gene felt great relief, although it was commonly tinged with guilt. But for those found to carry the abnormal version of the gene who learned they were certain to suffer the disease's calamitous end, the consequences of seeing the future were much more complex. Because there wasn't any treatment to slow the deadly damage caused by the mutant gene, scientists found themselves in the tragically awkward position of telling people they faced a horrific, yet unalterable fate.

Still, many people who belonged to Huntington's families sought the test. As a result, by 1992 psychologists and genetics counselors who carried out the tests in research centers had developed guidelines for "presymptomatic testing" of "late-onset disorders for which there are no available means to treat or prevent disease."[1] They recommended that those receiving test information first be carefully evaluated to determine their "readiness" to receive gene results. The researchers also urged that if an individual chose to proceed with testing, extensive and long-term follow-up counseling should be offered.[2] Still, experience showed that "adhering to these guidelines is time-consuming and costly," and many "medical centers have deviated from strict adherence to them,"[3] as the researchers put it.

Over the course of several weeks following Susan's revelation, Weber, Collins, Calzone, and a few other members of the research team made several decisions based on gut instinct, common sense, and consultations with those few medical people, like the Huntington's testers, who had some experience in the fledgling practice of predictive medicine. Right off, the Michigan team agreed to notify as many in the family who could be assembled in one place at one time. "We couldn't see dribbling out information to the family one person at a time over months," Weber says. "We already saw the great anxiety caused by letting out just one piece of it to Susan." In a decision that later sparked intense debate among the researchers themselves and many others in the field, the Michigan group decided to encourage as many of the family members as was possible to come to a single group-counseling session at the university at the end of October. Once there, each adult individual's genetic destiny was to be revealed in a private encounter with a doctor and genetics counselor.

The research group decided they needed help and recruited Barbara Bowles Biesecker, a genetics counselor at the university. The molecular biologists back at the lab also told the team they wanted fresh DNA samples to run the linkage analysis again. "Everyone wanted to be doubly and triply certain the results were correct before we handed them out," Weber says.

Calzone agreed to travel to the family's home to retrieve new tissue and blood samples from thirty family members. She called Janet and told her about the researchers' idea for releasing the genetic data all at once. She also asked Janet to gather together the entire clan on a Saturday afternoon in early October so that Calzone could collect fresh DNA samples. Biesecker agreed to accompany Calzone to Janet's home. "I wanted to meet the family and do some assessments" of the type suggested by the Huntington's experience, Biesecker says. "I wanted to see how much they knew, how much they understood, and who in the family might not be emotionally prepared or able to hear what we had to tell." The family still really knew very little about genetics. The information they'd received up until then was sufficient to gain their consent as subjects in a basic research project. "But now they were facing presymptomatic genetic testing," Biesecker says. "That was a whole new ballgame."

In the car drive north to meet the family, Calzone and Biesecker hashed out preliminary logistics for the mass family counseling session scheduled three weeks later at the university hospital. "Most of the ideas about how to structure the day [in late October] came to us during the drive," Biesecker says. But once at Janet's home, the two women got their first inklings of how little they could control the coming events.

During the late morning and into the afternoon, many of Dolly's and Mattie's siblings, children, cousins, nieces, and nephews streamed in and out of the house. Calzone took blood

from the family while Biesecker engaged each donor in conversation. Watching them interact, Biesecker was struck by the warmth
and closeness that the family members expressed for one another.
There was a lot of good-natured teasing between the cousins and
across the generations.

Biesecker had yet to meet Susan. But she looked forward to
meeting the woman whose close call in August was fast-becoming
a legend among the Michigan researchers. "I saw a woman walk
into the room and right away I thought, 'This has got to be Susan,'" Biesecker recalls, adding that seeing her "filled me with
shivers." Susan was wearing a white T-shirt tucked into a pair of
tight blue jeans. She was gregarious and funny, engaging most
everyone in the room. "This was a woman who felt comfortable
with herself," Biesecker recalls. Kathy Calzone says she noticed a
difference, too. At previous meetings, Susan had worn baggy pants
and loose-fitting tops. "This was obviously a new look," Calzone
says.

Unknown to Biesecker, she was carrying with her some chilling
news. To Biesecker, it was merely unalterable bits of genetic law
she felt obligated to pass along. But to the family, it was a genetic
bombshell. All along, Janet, Susan, and the other family members
believed that if a gene were at work, it only could be passed from
mother to daughter. That seemed right because only women were
getting cancer.

But during her "educational" talk that day, Biesecker stunned
those gathered with a fact of nature. First off, she told them the
laboratory scientists had determined from tracking the marker
that the chromosome carrying the tainted gene was a genetic
legacy passed down from Edward Stone and not from his wife. Indeed, some of the gene researchers elsewhere had a hint of evidence that men who received the aberrant gene had a higher than
normal risk of prostate cancer, the disease that killed the elder
Stone. Biesecker also told the women that all five other siblings of
Dolly and Mattie, including their four living brothers, also may
have inherited the same gene-carrying chromosome from their father. In turn, those five siblings may have passed the gene on to

their children and grandchildren. All told, between the siblings, aunts, uncles, and first cousins, there were thirty-five people who were at risk of carrying the dreaded gene.

Janet and Susan had two brothers sitting in the house who suddenly were drawn more intimately into the family nightmare than they had thought possible. Robert, 39 years old, didn't take the news well. Moreover, he became angry when Biesecker told him that she and Calzone didn't believe his 19-year-old daughter, Julie, seemed mature enough to hear results of a gene test, no matter which way it went. "He demanded that Julie be allowed to get her results," Biesecker says. As a result of numerous conversations with Robert in the following months, Biesecker learned more about the source of his outburst. "He was totally freaked out by the idea he might have this gene and he might have passed it to his daughter," she says. "All along he was dealing with this as his sisters' problem. In an instant, he learned it might also be his. It might not have been the best way to tell him."

Among those present were several grown daughters of Dolly's brother, Douglas. They, too, heard for the first time that males in the family were at risk for carrying the gene and passing it on. "I went to Janet's house that day to give blood and, you know, give whatever we could to help with the research, to help Janet and Susan and their families," says one of Douglas's daughters, Anna, who also was 39 years old at the time. "We never thought we were in danger. I still wasn't worried; there was no cancer at all in our side of the family. But it was a shock to hear we were at risk. I just couldn't believe it. We all got very upset. My dad was worried. We all were worried."

The concern among Douglas's seven daughters intensified the following week, when Biesecker called them to ask if there branch of the family would also come to Ann Arbor for a counseling session and to hear their test results. The women agreed to go the weekend after the one scheduled for Mattie's and Dolly's family. "It made me anxious," say Anna. "I wondered why they wanted us to come but I didn't want to ask too much. I thought we were safe."

Finally, Biesecker told those gathered that accumulating evidence from studying other families suggested that women with the altered gene had a higher-than-normal rate of ovarian cancer. "Nobody had told us that before," says Janet. "I thought, 'Geez, do I have to have my ovaries removed, too? When is the bad news going to stop?' "

For the next few weeks, the family and the Michigan researchers spoke frequently over the phone as they pinned down details of the group's counseling day. Back at the lab, bench scientists testing the family's fresh DNA ran into a minor problem analyzing the results of Mattie and her family. The lab scientists told Calzone they needed additional blood samples from Mattie. When Calzone called, Mattie told her she was leaving that day by car for a long-needed vacation with her husband, Frank. Calzone was frantic, and convinced Mattie and her husband to meet with her halfway between the university and their home, in the parking lot of a strip mall off a highway exit.

"It was pretty funny," Mattie says. "I sat in the backseat of our car and Kathy rolled up my sleeve, put in that big needle and took my blood right there in the parking lot." Adds her husband, Frank, laughing, "I kept waiting for a police car to drive up and arrest us. It looked like they were taking [illegal] drugs. It's one of the few things about all this that we can sometimes joke about."

The day before Halloween, the Stone family arrived at the oncology clinic at the university hospital to meet the new-age fortune-tellers. "The staff was very apprehensive," says Barb Weber. "I've been in the situation hundreds of times where I had difficult news to tell a cancer patient. But none of us had done anything like this. Francis [Collins] especially had a sense the world was watching. He told the family someone from the national media might be interested in their story. So we were nervous. We wanted to do it right for the family's sake, and also because we knew we were being watched."[4]

The most heated debate among the researchers centered on exactly what they should tell women who were found to harbor the abnormal gene. In particular, they were uncertain about whether

to recommend vigilant monitoring or prophylactic surgeries to remove breasts and ovaries. Henry Lynch's group in Nebraska that
had been collecting breast cancer families urged many women in
those families to have the surgery. But Mary-Claire King in California opposed the routine use of the surgical option, arguing
women should be counseled that being extra-attentive could
catch cancer when it was at its earliest and most treatable stage.
Moreover, "some women in our families who inherited the gene
weren't getting cancer at all, or were getting it very late in life,"
King says. "I didn't feel it was right or medically appropriate to
give women in their twenties or thirties a blanket recommendation to have their breasts and ovaries removed. It just wasn't
right."

By the time the family arrived, the researchers decided to take a
neutral position regarding prophylactic surgeries. The family sat
together in a large outer waiting room as the researchers called the
women and their closest family members into an offiice. There
they were met by a team that included one of the doctors, either
Collins or Weber, and either Calzone or Biesecker. Right off, the
researchers told the other sister of Dolly and Mattie that she and
her children had dodged the gene. "They breezed in and out," says
Janet. "They left right away and went shopping" in Ann Arbor.

The news wasn't so good for Dolly, however. She, of course, had
the gene. Moreover, she had given it to her daughter Janet and to
her son Robert, who had convinced the scientists to release the results of his 19-year-old daughter, Julie, by telling them that she was
older than she was. Julie, too, was told she had the gene.

Then things got messy. The researchers hadn't expected the
family to question one another as they came back to the waiting
room. But that's what they did. "We just didn't anticipate that
everyone would be sitting there ready to pounce on one another to
know their results," Calzone says.

Robert's daughter, Julie, didn't want to face the crowd. "Right
away with Julie we felt we'd done something wrong," says Calzone
who was in the room with the young woman. "She didn't want to
go back out into the waiting room. She didn't want to share her

results. We thought about having her leave by the back door, but, of course, that would have tipped everyone off. I realized right then that we'd taken away her right to privacy. I was very upset."

When the youngest of Dolly's five children, Scott, came out of his private session he told those who greeted him he hadn't inherited the gene. But he had. "My mother was getting a lot of bad news," says Scott, who had a baby daughter who now was at risk. "I was trying to protect her a little." He told his mother the truth a few days later.

Janet had expected bad news regarding her 18-year-old daughter, Meg. When it came, Janet was surprised by her anguish. "I felt so awful," Janet says. She walked into the waiting room, told Dolly the bad news, and said, "Mom, I feel so guilty for giving Meg this gene." Dolly looked at her daughter sadly. "How do you think I feel?" she asked.

"It hit me," says Janet. "I had passed the gene to my daughter, and I felt bad. But Mom had passed it to two daughters who had cancer, and now at least two granddaughters had [the gene], too. I could only imagine how she felt." Still, Janet was inconsolable. She went home that evening and cried. The next few days she had trouble getting out of bed. "Meg kept saying, 'Mom I can handle getting the gene, but I can't handle your reaction,' " Janet says. "It took me a long time to get over the news. I'm still more upset about it than Meg. She's young and feels strong. But I think about what she's facing. I want her to have mastectomies and have her ovaries out. But we don't know when she should do it. It's such a terrible thing for a young woman to face. I already was married and had my children when I had my breasts removed. It's such a different thing to ask of Meg."

The researchers also were shaken. Collins came out of one session especially upset. It was his job to tell Mattie's surviving daughter, Jesse, her test results. Jesse, of course, had removed her breasts five years earlier. But Collins told her she hadn't inherited the gene. Her mastectomies had been unnecessary. Jesse had wavered for weeks about whether to hear her results. Ultimately, she agreed to participate, mostly because she was concerned about her one-year-old daughter's future. The "good" news took a long time to digest. In the days that followed, she told her relatives she wasn't angry; that removing her breasts was "the best decision" she could have made at the time. Moreover, she was relieved for her children.

The day was especially difficult for Mattie. Pamela and Beth had between them a daughter and two sons who were at risk of carrying the gene. And two of Mattie's four sons (two didn't seek their results that day) had two daughters who were told they had the gene. "It was hard watching Mattie," says Biesecker. "It was like Russian roulette all day. This one has the gene; this one doesn't."

There was also some reason to rejoice. One of Lucy's two daughters was told she hadn't inherited the gene that killed her mother. (The other daughter was living in Europe and didn't attend the session or learn her results for several years.) "I was just so certain she had it," says Susan. "I couldn't believe how happy I was for her. We hugged and hugged. I couldn't believe it was possible." She adds: "I thought we'd be luckier than we were. I was surprised how many people did have the gene."

In the months that followed, the researchers argued over what had happened. Kathy Calzone was convinced that it was wrong to give the test results in such a public way. At first, Barb Biesecker agreed. "I would not have wanted to share the information with my family that way," says Biesecker. But as a result of counseling talks she held with family members over the next few months, she changed her mind. "I learned to appreciate how close the family was and how much support they provided one another," Biesecker says. "They were able to share the loss of those who had the gene and celebrate with those that didn't."

But in the days after the session, the researchers didn't have much time to react. They had to prepare for the following week when Douglas and his ten grown children were to descend upon them. Unknown to the family until they arrived at Ann Arbor the following Saturday, Douglas had inherited the gene. Three of his seven daughters, six of his ten children in all, had it, too. The researchers were bracing for an emotional day, especially since the members of Douglas's family had believed until two weeks earlier that they had escaped.

The researchers were fearful of how Douglas's family would take this sudden change in risk. But the research team wasn't prepared for what happened next.

Even after the researchers told Douglas's gathered clan, one of his daughters, Anna, a woman with bright red hair and a bubbly personality, felt hopeful about the family's chances. So when Barb Weber told her she had inherited the gene, "she was stunned," says Weber. Since she was only 39 years old, Anna never had received a mammogram. "We suggested she have one soon, and she said she wanted it right away, right there in the hospital," Weber says.

"I was in a fog," says Anna. "They had a nurse take me downstairs to have the test. Somewhere in the distance I could hear my mother crying."

All told, Douglas heard that day that six of his offspring were at risk of carrying the gene. "It was a goddamn nightmare," says Douglas, in a conversation several years later, his gruff voice belying a sweet soul. "I didn't leave 'em a whole lot to start with. To leave 'em this . . ." Tears well in his eyes and he turns away. "I can't even talk about it."

Concerned about Anna's mental state, Weber joined her at the hospital's imaging department. Weber told the technician that Anna was a member of the now-famous breast cancer family. The mammographer had already examined Anna's film and decided nothing was amiss. But alerted that Anna was in a family with a history of disease, the technician turned cautious. "[The mammographer] showed me the film," says Weber. "There was this very tiny whitish area that normally wouldn't raise any concern. We both decided to do the test again, a little differently."

The second mammogram was unmistakable. There definitely was a suspicous mass of tissue in one breast. "I felt like I was in a tunnel," remembers Anna. "I had trouble hearing what they were saying to me. I just wanted to go away. In the morning I was fine. Then they told me I had this risk of cancer, and then they told me

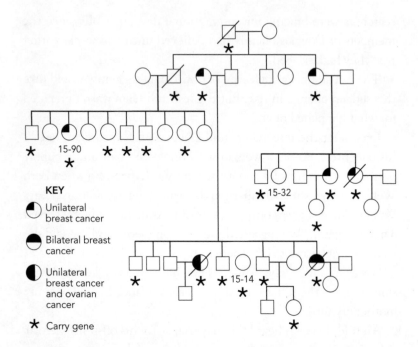

KEY

◐ Unilateral breast cancer

◒ Bilateral breast cancer

◖ Unilateral breast cancer and ovarian cancer

★ Carry gene

The figure above is a pedigree, or family tree, for Family 15. Men in the family are identified by the squares; women by circles. Those members of the family who carry the gene are identified with an asterisk. This pedigree has been slightly modified to protect the family.

I might even already have cancer. It was the most extraordinary thing. By the end of the day I could hardly hear what anyone was saying."

Two weeks later, Anna learned that, indeed, she had the very beginnings of cancer. The tumor was tiny and hadn't spread. Within weeks she had the breast removed. Says Weber, "The whole experience with Anna made me feel very weird. It was like a *Twilight Zone* episode. First we tell her she is likely to get breast cancer and then to catch it right there as a result of the test. On one hand, I felt terrible for her, and yet there was this feeling of almost morbid fascination with the power of the science. I still find it unbelievable that we could go from a [labcratory] reaction to predicting and catching cancer in a woman who believed she had no reason to feel at risk."

After Anna learned she had cancer, her younger sister, Adrienne, also a gene carrier, had a mammogram. As with Anna, the mammogram detected a potential problem. It was cancer, and within weeks, Adrienne also had her breast removed. "This gene thing is terrible," says Mariann, another of Douglas's daughters who carries the gene. "But the test is wonderful. We thank God that Janet and Susan pushed to have the test done." Adds Anna, "If Susan hadn't gotten results I would be dying of cancer."

Several years later, Mary-Claire King was asked her thoughts about the many women who might someday be affected by similar experiences. "Never in all the years I spent looking for the gene did I imagine researchers would use the gene discovery to make predictions about people's likelihood of getting cancer," she said. "We sought the gene to provide us a key to someday curing cancer. Maybe I should have thought about it more. But during all those years of looking for it, I wasn't ever really certain there was a gene. What's happened is as surprising to me as to anyone else."

6: The Game

By December 1993, just three years removed from Cincinnati, Mary-Claire King had become possibly the world's most influential woman scientist. Picked as one of *Glamour* magazine's twelve "Women of the Year," along with Supreme Court Justice Ruth Bader Ginsburg, author Nora Ephron, and Attorney General Janet Reno. She sat on prominent scientific policy panels, helped decide the fate of dozens of federal research grant requests, and was profiled in *The New York Times, The Wall Street Journal,* and numerous other national print and broadcast media. When James Watson stepped down as head of the National Center for Human Genome Research, the institute overseeing the U.S. effort to decipher the information contained in human DNA, it surprised no one that King was one of two finalists interviewed for the job.[1]

But her fixation on the breast cancer gene was taking a personal toll. The gene was an especially elusive quarry, much more difficult to track down than anyone had expected. Time and again, King and her lab colleagues believed they had the gene within their grasp, and each time it slipped away. King's hair turned gray as she watched hot leads evaporate and promising research avenues turn into useless dead ends. To friends and family she remained cheerful and enthusiastic. In media interviews she was an increasingly eloquent advocate for cancer research, and for more money for the human gene hunt. The war on breast cancer had become front-page news; Congress and the National Institutes of Health were under intense pressure from women's groups to pump more funds into the effort. In this environment, many began to view the breast cancer gene as science's best chance for unlocking the disease's terrible mysteries. Mary-Claire King was the embodiment of this scramble to capture the gene; a feminist scientist leading the charge against the ultimate killer of women.

But the pressure to succeed on such a grand scale and with all the world watching was beginning to show. Soon after the collaboration with Collins began, King had several tense exchanges over the phone with Barbara Weber, whom Collins had designated as his chief lieutenant in the project. The quarrel, says Weber, began over how the labs would divide certain technical tasks and then "disintegrated into disagreements over everything."

Collins chalks up the dispute to stress and King's previous lone-wolf style of research. "She'd run her own show all those years," Collins says. "Now we had to do a lot of sharing and trusting. Working out all the details was very hard for her and for us. Every conversation I had with Mary-Claire was a lengthy, twisting thing that often lasted an hour or more. And we talked every day. It was exhausting for both of us."

King was working furiously now to learn new molecular biology techniques, and always felt she and her lab had to work harder to stay even. The lab was toiling seven days a week, logging 10 to 16 hours a day. It wasn't uncommon for the graduate students to bump into one another coming or going from the lab at two o'clock in the morning to complete an experiment begun earlier in the day, or to get another going.

Moreover, the pursuit of the gene no longer was King's private, lonely crusade, but an international contest. Soon after the 1991 London meeting where Gilbert Lenoir confirmed King's original claim, a half-dozen top molecular biologists including Francis Collins, King's longtime nemesis a geneticist her same age named Mark Skolnick, Ellen Solomon, Lenoir, and others agreed to forge

an extraordinary alliance to quickly corner the gene. Many researchers hailed the joint effort, called the International Breast
Cancer Linkage Consortium, as a model for future gene hunts. It
was precisely the kind of coming together of great brains and resources needed if the cancer was ever to be bested. But by 1993,
this grand collaboration existed no more. Cooperation had been
replaced by competition and a personal rivalry that soon took on
the flavor of combat.

The scientists, many of whom are longtime friends who had
supported one another on other projects, acknowledge that they
felt compelled to contest for an incalculably rich genetic jackpot.
As the race became more protracted, and more scientists joined
in, it soon became clear to the contestants that great honor and
fame would go to the one who crossed the finish line first. Some of
the scientists readily acknowledged the stakes involved in what
some conceded was the greatest gene hunt ever.

Privately, the researchers began talking about the possibility of
a Nobel Prize for the scientist who finally fished the gene out of
the human gene pool. "Whoever wins will be one of the great he-
roes of science," said Mark Skolnick at the University of Utah. In
Great Britain, Bruce Ponder at the University of Cambridge said
finding the gene will "make a career. It matters hell of a lot."[2]

It matters so much because momentous discoveries are the life-
blood of the university-based scientist. Academic researchers, of
course, earn a great deal less than scientists in private industry. But
their rewards for publishing a significant scientific advance can be
substantial. Accolades and fame improve the chances of winning
research funding from the National Institutes of Health and other
government granting agencies. While scientists generally receive
livable salaries from their universities—and notable researchers
are even paid handsomely—their laboratories and the stipends for
graduate students who staff the labs are almost always at the mercy
of government funds or grants awarded by research-based private
corporations. In the early 1990s, competition for Uncle Sam's
beneficence to fund gene-related research grew ever more stiff.
Congress slowed the federal growth in science funding and more

graduate students interested in science turned to the increasingly popular field of molecular biology.

As a result, many university-based biologists warmly greeted the battalion of Wall Street financiers who were building hundreds of new "biotechnology start-ups" based on the gene-hunting scientists' technical wizardry. Faced with a government cash squeeze and the potential for significant earnings outside academia, many contemporaries of Collins and King succumbed to the siren call of industry. Many even began courting the venture capitalists, pitching them notions about new ways to attack poorly treated illnesses such as cancer with new gene-based technologies. Indeed, where the academic research community once sneered at colleagues who went "private," many now accepted the move as a legitimate strategy for financing research.

By 1993, several gene-hunters decided to follow that route, most notably Mark Skolnick, who for almost two decades had kept his gene-sleuthing investigations afloat at the University of Utah in Salt Lake City with a patchwork of government and private industry grants. Although Skolnick and his Utah colleagues had been searching for a breast cancer gene on and off nearly as long as had Mary-Claire King, he was stunned in 1991 when his application for federal funds to support his effort was initially rejected by an advisory panel of scientists at the National Institutes of Health.

Desperate to keep his pursuit of the gene alive, Skolnick formed Myriad Genetics, Inc., one of the first private companies whose sole goal was to find previously hidden genes that underlie common diseases. Armed with $10 million that he and his partners solicited from a few dozen private shareholders and from Eli Lilly & Co., the large drug-maker, Skolnick was able to build a private gene-hunting lab in Salt Lake City that within a few years rivaled any in its size and staff.

Even as academic colleagues embraced the financial support of biotechnology, Myriad's entry in the breast cancer gene race angered and frightened the other breast cancer gene contestants. "I worry what Myriad will do if it finds the gene," said Ponder of Great Britain. "The investors will want a profit from the discovery and that could pose all sorts of problems we've really not faced before in the field." Some even privately acknowledged their envy. "If Mark's company finds the gene he stands to earn millions of dollars," one of the researchers said. "It doesn't seem right to make money that way."

The breast cancer gene search, in fact, was now one of the dozens of enterprises fueling a new commercialization of biology whose promise of new medicines and gigantic profits also produced new ethical conflicts and quandaries. Advances in understanding the workings of the cell at the most fundamental levels had reached a point where it was possible to transfer the science from academic labs to commercial ones. Scientists now saw they might create totally new types of medicines by synthesizing chemicals to interrupt previously hidden disease pathways. At the center of these new pathways was the disease-causing gene.

"Suddenly, biologists who were skilled at isolating genes realized they were producing completely new targets and new strategies drug companies could use for intervening in diseases, such as cancer, that still eluded adequate therapy," says Stephen Friend, a Harvard University molecular biologist.[3] "Each new gene or set of genes opened a scientific door to these unknown disease pathways and targets."

The genes and the people who find them, therefore, were becoming extraordinarily valuable. But to be a player you had to make the discoveries. The value of the find rose if your lab made the find by itself, with as few collaborators as possible to dilute its value. "It's human nature to fight over something as precious as uncovering a human gene," said Ray White, a top molecular biologist who is the captain of a competing team of gene hunters a few blocks away from Skolnick's labs at the University of Utah. "And right now the most precious of these is the breast cancer gene."

"It would be wonderful if everyone agreed to share [research]," said Ellen Solomon of Great Britain's International Cancer Research Fund in London. "But scientific inquiry has always had an element of competition. And these days, there is no question that natural competition is intensified by the financial pressures we all feel. Competing against each other like this has been very unpleasant but it appears to be what we have to do."

"The idea that at any moment someone will find the gene and that we have as good a chance as anyone in the world keeps me going," says 28-year-old Lori Friedman, one of three principal graduate students in King's lab.

King was repelled and transfixed by the turn of events. Sitting in her large, brightly lit, but run-down fifth-floor office on the Berkeley campus in early 1994, King complained that the competition's bare-knuckle battles cheapened the research's goals. "Calling it a contest or a race—and some of us are calling it that—is just an awful way to describe what we're doing, especially when you think about the women who are depending on us. I don't mind this being called a race, if you mean a race against the disease. Every time we hear of another woman who died of breast cancer we take it personally. We should have the gene by now. I've really internalized the need to get this done. But I have to play by the rules. We're a competitive society, and this is the way these things are done. In fact, I guess, this is the way most important things get done."

The quest provided King an enviable platform for her progressive and feminist ideas. In the *Glamour* magazine interview, for instance, she bitterly attacked the insurance establishment for failing to provide adequate medical coverage for many women with the disease. "This is no way to run a country," she told the magazine's interviewer, referring to the fact that patients without insurance aren't getting proper care. "Genetics is only a small part of the problem [of breast cancer]," she said. "Health-care reform is essential to our survival as women."[4] The statement was typical of a brash honesty rare among scientists beholden to the goodwill of the federal government and private foundations. But few of her colleagues had the opportunity or audience she now had.

Equally important to King was the almost uniformly generous outpouring of public accolades she was receiving from her contemporaries. "I have to sit back in amazement when I think about her decade-long effort to collect families and then run [DNA] marker after marker for years without much evidence or support that the technique would yield results," White said. "It's a real testament to her strong-minded and focused personality, especially since she really had to learn the technical side of gene-hunting on the fly. She was like a terrier with a bone. Whoever finally isolates the gene will be rightly hailed. But Mary-Claire deserves great credit for what she accomplished. It's helped to transform the way we think about and carry out cancer research."

7: Cancer Families

I t was a childhood incident that had drawn Mary-Claire King to her life's work. She had experienced the tragic sting of cancer at an unusually young age. Her best friend from the age of 4, a girl who lived nearby, was struck by a childhood kidney cancer called Wilms's tumor before she was 10 years old. The friend died in early adolescence after a difficult and lengthy battle. "The two girls were very close, and Mary-Claire was at her home a great deal and suffered as she watched her go downhill," King's mother recalls. "I know for a fact that Mary-Claire never got over that."

King graduated from Carlton College in Minnesota with a degree in mathematics in 1966, then began graduate school at Berkeley. The greatest single influence on her professional life came the day she met Allan Wilson, a 32-year-old biochemist. In a campus filled with charismatic radicals, King found no one more compelling than Wilson, a gentle and whimsical New Zealander whose shy manner hid the soul of an intellectual revolutionary. Wilson had arrived at the campus three years earlier from Brandeis University and was just beginning to stir up a commotion with one of the most controversial ideas about human evolution since Darwin.

Wilson believed DNA could be used as a "molecular clock" by which scientists could literally go back in time and identify exactly where and how humans originated. At the time, all ideas about human evolution came from uncovering and studying fossils of the earliest humans and humanlike primates. But beginning in the mid-1960s, Wilson conceived the idea of comparing DNA sequences of genes of different human races and primate species to find the time and place where humans first split from apes. In a fury of technical discoveries, Wilson had created a genetic timepiece from which he believed he could look back in time and determine the earliest biological events.

By the 1970s, Wilson was using his time-traveling technology to challenge the orthodox view that humans emerged from primates 15 million years ago, arguing that 5 million years back was more accurate. And in 1987, Wilson's timepiece caused an international uproar when the scientist and two colleagues theorized that all humans have a common ancestor among a tiny population who lived in Africa 200,000 years ago. In a move that brought him instant celebrity, Wilson said all humans descended from one woman in this group. He called her "Eve," the mother of all humans.

The claim kicked up a storm of criticism and charges of scientific sensationalism. But the technique of using variations in DNA sequences to identify human origins caught on, and other researchers began using it to piece together historical migrations of people, thereby identifying, for instance, the Asian origins of specific American Indian tribes. (One scientific team has used the Wilson technique to trace the ancestry of most American Indians to a small band of people that migrated across the Bering Strait 15,000 to 30,000 years ago. King was later to use a similar method to track down the children kidnapped by the Argentine junta.)

Back in 1967 while Wilson was first tinkering with his genetic clock, King and he met at "one of the many [meetings] in the 1960s' endless effort to make both science and the University more relevant to the needs of the world,"[1] King later wrote in a personal letter. The two became friends and Wilson, who was passionate about genetics, persuaded King that through inquiries into the most basic nature of life she could make a difference—much more than was possible in mathematics. "I wanted to do something with an obvious environmental impact," she recalls. But she soon grew impatient with the research and was unable to see a connection between her studies in bacterial genetics and the political winds blowing across the land. King dropped out of school and joined consumer activist Ralph Nader's California Project, an investigation into environmental issues. She even worked personally with Nader, who offered her a plum staff position.

Before she took it she spoke to Wilson who convinced her to give genetics another try. But the roiling times, and the compulsion to participate in the activism, remained a difficult distraction for even the most focused of young people. King went through several more years of personal experimentation before giving in to Wilson. She went abroad to study first in Edinburgh, Scotland, and then in the Soviet Union. In 1970, she returned to Berkeley uncertain how best to proceed in a career in science. She was 24, had spent four years in and out of graduate school, and "I only had two failed projects to show for it," she says.

Finally, King dragged herself despondently to Wilson. "I was completely discouraged by the failure of every experiment I attempted in the lab in which I was then working and exhausted by the apparent futility of our grassroots political work," King wrote years later.[2] "I was planning to leave school [again]. I went to Allan's office . . . and poured out all my discouragement."

"Allan listened to all this, smiled his bemused little smile and said, 'If all of us whose experiments nearly always failed stopped doing science, no one would be left. I certainly wouldn't be. We need to think of a project that will be interesting whether you get a positive result or not.' Of course, I was completely at a loss to understand what he was talking about, but what he had in mind was a terrific idea that became my Ph.D. dissertation." The idea, to compare the genetics of monkeys and humans, was to serve years later as the scientific foundation upon which King built her earliest strategy for going after the breast cancer gene. But at the time, King was pleased simply to embrace a meaningful and exciting project.

Wilson's attitude about scientific inquiry was that above all else it should push, provocatively if necessary, the limits of knowledge. Over the next few years, King drew courage and inspiration from Wilson's scientific anthems, using them time and again to spur her on. In 1975, under Wilson's tutelage, she published his suggested Ph.D. thesis in, of all places, *Science* magazine. Most astonishingly, especially to King, the research was showcased as the magazine's cover story, illustrated famously with the drawing of a chimpanzee.

The *Science* article argued that humans and chimpanzees aren't merely genetic cousins, as is often claimed, but, instead, are almost genetic look-alikes. The paper, titled "Evolution at Two Levels in Humans and Chimpanzees,"[3] and coauthored by King and Wilson, describes a series of experiments in which the researchers compared gene sequences of many proteins used by both species to carry out similar life-sustaining tasks. To their surprise, King and Wilson found the average human protein is "more than 99% identical to its chimpanzee counterparts." In other words, people and apes are nearly as alike at the genetic level as are one human and another.

How, then, the authors asked, can science account for the significant differences in anatomy, behavior, and intelligence between monkey and man? King and Wilson suggested that, as with many things in life, it all depends on timing. That is, the distinctions between the two species occur when strikingly similar genes are turned on and off at different moments during the first days and months of life. In other words, humans evolved differently from the chimp as a result of natural changes, or mutations, that caused differences to just a tiny few bits of genetic material that dictate, much like a traffic light alternating the flow of automobiles in a busy intersection, how and when most genes carry out their tasks. The proximity of the two species to each other, genetically speaking, led Wilson to suggest that humans and the chimpanzee may have diverged from a common ancestor a mere 5 million to 6 million years ago, not 30 million years in the past as paleontologists had believed. (That claim is now accepted among anthropologists.)

As might be expected, the suggestion that man and chimp are genetic siblings—indeed, almost twins—sparked quite a scientific row. Within weeks of its publication, *Newsweek* magazine, among others, took notice of it. In a feature article in its science section, the national newsweekly said the findings of Wilson and the 29-year-old King buttressed the emerging research of a small group of controversial biologists whose new laboratory methods of gene analysis had led them to begin "furiously debating Darwin's principles" of evolution.[4]

Nonetheless, the row gave King a taste for fame and the opportunities that can come to those whose work is noticed. It was something she often thought about in the years to come when she struggled in frustration and obscurity. Many years later King predicted, only half-jokingly, to the esteemed editor of the journal *Science* that someday she again would produce research prominently displayed on the magazine's cover. But next time, she hoped, it would be for the discovery of the breast cancer gene.

Mary-Claire King remained close to Allan Wilson for years until he died in 1991 of leukemia, and by then, he was no longer her mentor, but more like a beloved "older brother." The publication of the *Science* paper marked the end of their professional collaboration. By the time it came out, King's restless craving for a meaningful project of her own—as well as the belief that she might never make a significant contribution as a laboratory scientist—led her into a different field of genetics.

In 1972, King met and married Robert Colwell, a zoologist at Berkeley, who shared her desire for social relevance in science. In 1973, Colwell received a Ford Foundation grant to study ecology at the University of Chile, and King went along to teach genetics. Within a few months of their arrival, Chilean President Salvador Allende's socialist experiment was shut down. Allende's government fell and Allende himself died in a right-wing coup that was supported by the U.S. Central Intelligence Agency.

"I had this great hope that in Chile I'd finally do something visibly constructive, as far as helping people," King says. "But after the coup, everything in Chile changed." Disillusioned and de-

pressed, she and Colwell returned to Berkeley. In need of a job, she
reluctantly took one in the seemingly unrelated field of cancer
epidemiology at the University of California, San Francisco. Although she didn't know it at the time, she fell into research that
was to consume her.

Epidemiology is the Sherlock Holmes branch of medical research. Its practitioners are detectives who try to "solve" the
causes of illness by searching for disparate clues among large
groups of people. For instance, epidemiological studies first uncovered the association between smoking cigarettes and lung cancer
by surveying cancer victims and compiling lists of behaviors common among them. It is epidemiologists who are called in when
there is any mysterious outbreak of an infectious disease, such as
the strange and lethal flulike disorder that struck down a group of
middle-aged men attending an American Legion convention in
Pennsylvania in 1976. The researchers finally tracked the cause
down to a bacterium that can contaminate air-cooling systems,
isolating and naming the microbe as the cause of "Legionnaires'
disease." Epidemiologists helped identify the common risk factors
among the first patients with AIDS, showing that the disease was
being transmitted through shared blood, and eventually helping to
isolate the culprit as a virus transported in the blood supply.

By 1974, when King was hired by a UCSF team of epidemiologists, cancer ranked second to heart disease in the murderous path
it was cutting through American lives. Politicians and public
health officials were baffled and alarmed by the seemingly rapid
increase in the rising incidence of both diseases among Americans. Some health officials were warning that Americans in untold numbers were being threatened by a twin epidemic, each as
puzzling to contemporary scientists as was any of the mysterious
disease outbreaks that had endangered humankind in times past.
It fell, therefore, to epidemiologists to unravel the causes for these
scourges.

By then, epidemiologists were beginning to pick apart the many
complicated "risk factors" that increase one's likelihood of contracting a heart-related illness. The epidemiologists had found a

powerful link between consumption of high-fat foods and heart attacks. It would be several years before scientists understood exactly how dietary fats such as those found in animal meats were converted into the waxy plaque that can disastrously block blood flow in the arteries leading to the heart. But the link found by epidemiologists between eating fat and coronary heart disease was statistically so profound doctors began warning Americans to sharply restrict the use of dairy and animal fats such as butter, steaks, ice cream, and certain cooking oils.

Epidemiological researchers surveying large populations of patients found family history was consistently a major risk factor in heart disease. Certain people could gorge themselves on fatty foods without any untoward effects, while for others this kind of diet was literally a death warrant. The difference, scientists surmised, was in the inheritance of an unknown set of genes. But scientists were unable to find any similar overpowering environmental agent or lifestyle to tie to the many different types of cancer. Except for tobacco use, which was closely associated with lung cancer, cancer epidemiologists were stumped in their search for a "smoking gun" to blame for the vast majority of other cancer killers, such as cancers of the breast and colon, or the leukemias, which are cancers of white or red blood cells.

King was enlisted by a research team at UCSF interested in using her genetics background to help begin to disentangle the influence of inheritance and "environmental" factors in breast cancer. The group screened large numbers of breast cancer patients, trying to find what they had in common. "We looked at thousands and thousands of questionnaires," King says. "I was seeing double, trying to categorize all the possible risk factors" uncovered in the surveys.

After working on the project for a while, "I was struck by something that others had seen before and that seemed so obvious I couldn't think of anything else," King recalls. "In so many women, the only shared risk was that some very close relative—a sister or mother or grandmother—also had breast cancer. Other factors—age of first pregnancy, consumption of dietary fats—also were present. But given my background [in genetics], it's understandable that I homed in on inheritance."

King didn't realize it at first, but she finally had found her cause. From her days as an activist and as a protégé of Allan Wilson she had learned the value of swimming against the established scientific tide. But only years later did she understand that she had chosen to be a maverick. At first, it merely seemed "to me as if I'd undertaken a very reasonable project," she says.

In fact, most human geneticists considered the search for cancer-causing genes as misguided heresy, a highly speculative waste of time. Instead, most established human genetics researchers were engrossed by the search for the cause of the 4,000 or so obviously "inherited diseases" that victimized humans. These are illnesses, such as muscular dystrophy or cystic fibrosis, with a clear-cut, undisputed genetic basis. Although scientists were able to discern powerful patterns of inheritance within a family that indicated a single gene was at the heart of one of these diseases, there was no way for researchers to look at the 23 pairs of human chromosomes inside the cells of an affected person and "see" exactly what tiny bit of DNA, or precisely what gene, was responsible for a particular genetic disease.

From time to time, researchers looking under a microscope could see damage to a particular region within one of the chromosomes. Perhaps they saw that a chromosome was shorter than usual because a section of DNA was mysteriously missing, as if some mischievous molecular scissors was at work. Or maybe they saw that a chromosome was stouter than normal because an extra bit of a neighboring chromosome had somehow become attached to its midsection.

Scientists don't know what causes these strange insertions and deletions, although they do know they are relatively rare occurrences that happen during the first few divisions of an embryo within the hours and days after a female's egg cell has been fertilized by a male sperm. Most chromosome anomalies are so destructive to the developing embryo and fetus they cause a natural miscarriage, often within the first days or weeks of life. But in many other cases, the damaged fetus proceeds to birth, albeit crippled in some form that will ultimately lead to health problems in the newborn child or, perhaps, later in life as the child matures into adolescence and adulthood. Down's syndrome, for instance, results from an extra bit of chromosome 21 somehow tagged onto the normal pair. The damage causes a range of mental and physical problems that, when especially severe, can be fatal in early adolescence.

Seeing these chromosomal alterations under a microscope sometimes helps scientists to "map" the location of the gene underlying an inherited disease. In other words, they can pinpoint the approximate site of a gene as residing somewhere in the altered section of a particular chromosome. But scientists had yet to figure out a way to actually identify the specific "damaged" gene at fault. That's because the site where gross chromosomal damage was found might be the home to dozens or even hundreds of genes whose identity was invisible to the most powerful of microscopes or gene-sleuthing techniques of the time.

Scientists believed by 1974 when King joined the staff at UCSF that there were about 50,000 to 100,000 human genes scattered across the vast genetic terrain within the boundaries of the 23 pairs of chromosomes nestled in the nucleus of every human cell. By identifying chromosome damage they had mapped a few hundred of these genes to a specific chromosome. And through a combination of tedious experiments, scientists had actually isolated the letter-by-letter base pair sequences, or chemical design, of only several dozen of these mapped genes.

Given the technology of the times, scientists restricted their gene-hunting to diseases clearly caused by the inheritance of a single gene. Any other gene hunts were regarded as beyond reach of the most technically adept. Researchers did believe genetics played a role in some common health problems such as diabetes or heart disease. How else can it be explained that some smokers develop lung cancer and others don't? Why do only some of the many people who consume high-fat diets develop heart attacks? Inheritance must be an important factor. But, scientists assumed, a susceptibility to these diseases was likely the result of inheriting several genes, not simply one gene as was the case, say, with cystic fibrosis or muscular dystrophy.

As for breast cancer, geneticists said there wasn't enough compelling evidence to indict one or more genes as conspiring to set off the terrible sequence of events underlying disease. Instead, scientists believed that breast cancer was mostly "acquired"—that is, caused by exposure to some unknown set of environmental agents during one's lifetime—and not inherited from genes handed down from one's parents. Familial clustering of the disease was due to happenstance, the scientists said. Because breast cancer was a relatively common illness it wasn't unusual, scientists said, for more than one family member to fall victim to it. Moreover, if the cause was some exogenous agent, such as lifetime exposure to an excess of fatty foods or an environmental chemical in the air or water supply, then it was likely that people in the same nuclear family would share this risk. Except for occasional claims to the contrary from a handful of scientific heretics and eccentrics who delighted in tweaking their more cautious colleagues, that was the dominant scientific view of breast cancer in the mid-1970s.

Even so, in 1976 Mary-Claire King persuaded officials at Berkeley's School of Public Health to bring her back across the Bay with a faculty appointment in the epidemiology department. Her task was to isolate, if at all possible, the role of genetics in common diseases such as hypertension and cancer. King was thrilled to be working again in Berkeley. She and Colwell now had a baby daughter, Emily, born in 1974, and they had a small house in the steeply rising hills above the Berkeley campus with a striking view of San Francisco across the Bay, the Golden Gate Bridge, and the open horizon beyond.

King returned to Berkeley armed with more than just a hunch. While at UCSF, King had come across the research articles of a handful of iconoclasts who claimed they were tracking a genetic link to breast cancer. Moreover, King believed that some of the lab techniques she and Wilson had used to compare chimpanzees and humans could be used to test these early genetic claims.

The very first recorded hints that breast cancer ran in families came in the nineteenth century from several remarkably obser-vant physicians. The most famous was a Parisian surgeon, Paul Broca, whose sharp intellect led him to discover, among other things, the brain's speech control center (known today as *Broca's area*). In 1866, Broca compiled and published an extensive list of women with the cancer in his wife's family going back four genera-tions.[5] Broca had no way of knowing what mysterious power was at work because the notion of inheritance and the natural laws that guide it had yet to be explained by a contemporary named Gregor Mendel.[6] To Broca, the cancer cluttering his wife's family tree was simply a phenomenon worth noting and pondering.

In the 1920s and 1930s, several other researchers used a variety of epidemiological methods to collect cancer-prone clans whose clustering of cancer cases was often unknown even to many mem-bers of the families. One research team in Denmark pored through a national registry of cancer cases, and plucked out 200 Danish families with multiple cases of breast cancer and other forms of cancer, too.[7] The study's most lasting contribution to science was that it produced an investigating technique that has become the standard for uncovering cancer families. Under Danish law, doc-tors back then, as they still must today, submitted to a national health agency the name of every patient with cancer.[8] While agreeing to keep the names confidential, the Danish researchers were allowed to contact the "proband," or specific cancer patient listed in the registry (or a close relative if the proband was de-ceased), for questions about the health status of other members of a family. In a stepwise investigation, the scientists then con-structed pedigrees when relatives could recall cancer cases among close and distant kinfolk. Each new case or relative was contacted and they, in turn, led the researchers on to other relatives. Medical records or the patients themselves were examined to determine the exact disease diagnosis.

Several research efforts in particular grabbed the attention of Mary-Claire King and Nicholas Petrakis, a cancer epidemiologist she first hooked up with at UCSF. In one study involving 510 families, researchers showed that breast cancer occurs with a significantly higher frequency in sisters and mothers of the probands than in the first-degree female relatives of similar women without cancer. In a report in 1948, this research team concluded that "a specific genetical agent, responsible for the disease, is postulated, which may be inherited mainly through the maternal line."[9]

Further evidence came from epidemiological surveys from David Anderson at M. D. Anderson Medical Center in Houston, whose analysis of a collection of cancer families was the first to show that the risk of breast cancer to first-degree relatives of patients who had cancer prior to menopause was three times higher than normal. From this same data, Anderson showed that if the patient had cancer in both breasts, the risk of cancer among close female relatives was five times higher than among women without a family history of the disease. The risk of cancer for close relatives jumped to nine times higher than normal if the proband had cancer before menopause and in both breasts.

"These women appeared to have in common something much more powerful than environment or lifestyle," King says. "We believed it was likely a gene or set of genes inherited by some relatives and not by others."

To chase after such a gene, King and her colleagues knew they needed to prove the cancers were popping up in family members in a statistically consistent pattern that followed true to Mendel's laws of inheritance. This required the collection of numerous large families in which the medical evidence of the breast cancers was well documented. Even if they found these families, the task ahead would still be daunting, for they then had to determine whether affected women in these cancer-prone families were inheriting the exact same stretch of DNA on a particular chromosome. Hidden somewhere on that bit of DNA, the researchers guessed, might be the culprit gene. But finding the families and then pinpointing the chromosome site wasn't going to be easy. "Back then, I didn't know what was possible and what wasn't," King says. "I just felt strongly that it was a project worth taking on." It was, she says, the long-sought "relevant" use of her scientific skill.

King and her colleagues knew that to dig up cancer-prone families of their own they would have to embark on a search similar to the one undertaken by the Danish researchers, an exceedingly difficult, expensive and time-consuming task. They soon heard about specialist Henry Lynch, whose cancer families would later prove to be valuable to Gilbert Lenoir.

In less than three years of looking, he had tracked down two large midwestern families consumed by cancer. In one family, Lynch reported finding evidence of 51 different cancers going back four generations. The other family was littered with 27 cancers in just two generations. While noting in a published report that other factors, such as a virus, might account for the phenomenon, he argued that it might also be caused by dominant-acting genes. "Intensive studies of 'cancer families' such as these may lead to new clues to the etiology of carcinoma," he boldly predicted in the report's published summary in 1966.[10]

Lynch's claim was viewed with deep skepticism by experts in cancer and genetics. Many believed cancer families, such as the two uncovered by Lynch, were freaks of nature with little value to mainstream medicine, and Lynch was largely ignored.

Still, one collection of midwestern families he published in 1972 that was studded with breast cancer caught King's eye.[11] "Lynch had collected these families much like an anthropologist investigating a strange culture," King says. "I contacted Lynch because he had a wealth of data. But it was clear he wasn't certain how to take the data further and map a gene."

The reason was that scientists lacked the tools to track most traits as they passed through a pedigree. But King believed the basic idea she and Allan Wilson had used to compare the genetic likeness of humans and chimps would serve as a model for hunting down the gene. This technique relied on the use of so-called polymorphic gene markers. The trouble for researchers in 1976 when King took on her mapping project was that scientists simply had yet to uncover many useful gene-mapping "markers." In fact, the first fourteen years of her project were dedicated principally to searching, largely in vain, for reliable markers.

A marker is some visible or detectable trait inherited along with an unknown gene. Eye color is a good example of a potential gene marker. If everyone in two or three generations of a family who developed a disease also had blue eyes, and those without disease all had green eyes, then one might conclude that the disease-causing gene was being inherited along with the gene for blue eyes. As a result, geneticists say the two genes are linked, meaning that the blue-eye gene can be a marker for the disease gene. If scientists know where the blue-eye gene is located in the genetic geography of DNA, then they can map the disease gene to the same site.

One of the human gene markers was found in the 1930s when color blindness, which is inherited, was linked to hemophilia, an inherited disease in which the blood fails to clot adequately. Scientists uncovered the link in a family where every male who had hemophilia also was color blind. So tightly linked were the two genes that in this family scientists could predict who had inherited the gene for hemophilia by simply identifying who was color blind.

But for a gene to be a useful marker it has to vary slightly in its size or structure from one person to another. This way, if a father has one form of the marker and a mother has another, geneticists can look at each offspring and tell from which parent the child inherited the gene that is linked to the marker. Because a genetic marker occurs in a variety of forms, or morphologies, researchers call it polymorphic. With a linked polymorphic marker, scientists can watch a gene as it flows through the generations. But when King took on her mapping project, few polymorphic traits were known, and fewer still were mapped to their approximate locations in the maze of DNA.

Despite the absence of markers, King was now thoroughly immersed in the quest. In 1980, King, her collaborator, Nick Petrakis, and Lynch believed they had finally found the much-sought-after marker. But it wasn't long before the discovery was shown to be an error.

Around the same time, on April Fool's Day, 1980, King's husband told her he was leaving her for another woman. "I remember it so well because in this same thirty-six-hour period I received news that I'd won tenure at Berkeley, Rob told me he was leaving, and our house was burglarized," King says. "Looking back, I don't think it was possible to be a good wife, a good mother and a good scientist. I just didn't have enough energy."

The breakup hurt King more than she first realized. "I was strong, or at least I felt I was strong at the time," King said in a conversation in 1994. "But all these years later it still hurts. It hurts a lot."

The combination of her personal problems and her gene-mapping error rocked King's belief that she might ever solve the breast cancer puzzle. "I now realize that for several years afterward I was numb," King says.

8: Gold Mine

A thousand miles away from Mary-Claire King, in Salt Lake City, Mark Skolnick was on a remarkably similar crusade. Like King, Skolnick was convinced, despite prevailing medical dogma, that a breast cancer gene existed. The proof, he firmly believed, lay in the high Utah countryside where, during the 1970s, he dug up evidence of unusually high rates of breast cancer coursing through several large families descended from Mormon pioneers who settled in the area more than 100 years earlier. Like King, Skolnick believed that by analyzing the families he might someday solve the breast cancer riddle. And, like Mary-Claire King, by the early 1980s, Mark Skolnick was baffled as to how best to proceed.

The two scientists, in fact, shared many similarities. They were about the same age and their paths had crossed several times, sometimes without their knowing it. They were both pursuing graduate studies in 1969 at the Berkeley campus of the University of California, although in different academic departments. They considered themselves gifted problem solvers, and their skill interpreting gargantuan mounds of data led them each, through twisting careers routes, to the field of population genetics. Their world view—progressive and committed to social change—was shaped by the civil rights movement and the Vietnam War. Both were determined to make a significant contribution to society, both wanted attention and praise, and both were convinced they could satisfy their personal and professional goals by cracking the mystery of cancer.

They agreed on one other thing, too. By the middle of the 1980s, Mary-Claire King and Mark Skolnick believed they had good reasons for not liking each other. They had briefly shared a federal grant in the late 1970s, and as sometimes happens as ambitious young scientists grapple for scarce financing, the two researchers feuded over the grant's control. When, soon afterward, Skolnick criticized a marker claim by King, the ill will between them heightened. The two became well-known rivals within the small genetics community.

Still, despite their mutual dislike, their shared quest bound them like two strands of DNA. And in the years to come, their lives and research would intersect many times, culminating in 1994 in a grand and surprising finish to the search for the breast cancer gene.

It is ironic, therefore, that it was Skolnick who found the solution to King's marker problem. The find suddenly opened the way for molecular biologists to create hundreds and hundreds of "polymorphic" gene markers sprinkled throughout the genome.

In 1978, Skolnick and three other scientists made a major advance in marker technology, and soon won international acclaim. "There is no way to overstate the importance of the [marker] discovery," says Francis Collins.

The genetic markers that Skolnick and his colleagues identified, much like landmarks in a barren Utah desert, pointed scientists to the location of the hidden genes that directly trigger incurable inherited disorders such as Huntington's disease, cystic fibrosis, and muscular dystrophy. The same markers led to the creation of the new forensic discipline of DNA fingerprinting used to convict criminals and to help free many others wrongly charged with crimes. It made possible early diagnostic testing of genetic disorders in developing embryos and fetuses. And it eventually persuaded biologists that the tools were at hand to undertake the so-called Human Genome Project, the ambitious global effort to identify all 100,000 or so human genes—the set of instructions for human biology.

As important, the markers made it possible for a small group of scientists, among them King and Skolnick, to launch an earnest bid to isolate the genetic component of widespread diseases not considered genetic in origin, but for which the real cause remains unknown. While many sophisticated scientists concede that inheritance—one's genetic constitution—somehow places some people at high risk of developing certain diseases, exactly how powerful a role genes play in disease formation remains controversial.

Still, one need only observe the many families where a father and several of his sons are stricken with a heart attack to suspect a conspiracy of inborn forces. It is reasonable, therefore, to assume that something within a person, some part of his or her genetic blueprint, makes him or her fit to withstand exposure to "unhealthy" lifestyles. But without firm proof of the existence and nature of such a set of inborn biological factors, there is no way to say with scientific certainty that heart disease, cancer, mental illness, or alcoholism truly "runs in the family."

The proof, of course, would come only from the identification of a gene or genes that, in various forms, make one able to withstand or wither in the face of environmental factors such as diet, air pollution, or lack of exercise. The markers supplied, finally, the crucial missing technology that dreamers like King and Skolnick, and even Henry Lynch, needed if they were ever to distinguish the specific inherited and environmental ingredients that together create the symptoms that are the hallmark of the common and incurable diseases that plague mankind.

It seems reasonable then that Skolnick's participation in the marker discovery should have assured him financial and professional prosperity. But it didn't. Although he was made a tenured professor in genetics at the University of Utah, by 1985 Skolnick found himself on the losing end of a nasty turf war with a powerful colleague and former collaborator over who would control the

University of Utah's genetics department. Just about when Mary-Claire King was reenergizing her breast cancer gene search, Skolnick was having difficulty landing federal research grants large enough to sustain a similar assault of his own. He had only a tiny laboratory and a small research staff shunted off the university's main campus and onto an adjoining, quiet industrial park.

As a result, he and his family retreated to a suburb outside New York City, where he went on a one-year sabbatical, tail between his legs. He was angry and depressed. "I needed to get away and figure out what to do," Skolnick says. "I was convinced that [in Utah] I'd helped uncover the best resource on the planet for getting at the underlying genetic cause of cancer and other common illnesses, including heart disease, diabetes, and even such things as obesity. I just didn't have the resources to take advantage of it."

As Skolnick often pointed out to himself and others, however, he still owned something that was the envy of the world of human genetics. He held key access to the Utah genealogy, an especially rich collection of extensive families he helped identify. The genealogy contained numerous families like those uncovered by the Dutch doctors years before and by Henry Lynch in the 1960s. It provided Skolnick with a "living laboratory" in which to observe common maladies as they snaked through a family's several generations. The Utah families were especially informative in this way because they were Mormons. By tradition Mormon couples typically produce large families with about eight or nine children. If a gene were the seed that over a lifetime of environmental watering grew into a disease such as diabetes, arthritis, high blood pressure, or cancer, its presence was likely to be observed only in families with many siblings, where there was an opportunity for such gene-based diseases to surface as familial clusterings. The Mormons also made a religious practice of keeping track of their ancestry, and, as important, most relatives lived in Utah. That made it feasible for researchers to identify and track down family members.

It is, in fact, the Utah genealogy that made it possible for Skolnick and his colleagues to invent the revolutionary gene marker technology. The genealogy was the gene hunters' gold mine.

"When I first arrived in Utah [in 1974] I couldn't believe my good fortune," says Skolnick. "I knew I'd found my life's work." But, he adds, "there wasn't any way, of course, that I could have guessed how important" that work would be.

Skolnick is thoughtful, articulate, and confident. Words and ideas ("big ideas," he likes to say) spring from him with seeming ease as he strides across subjects as varied as the origin of man, computer software, Italian cooking, or high-country snowboarding. Slightly under six feet tall, with a gentle paunch, his straight dark hair and mustache cared for in only the most casual way, and uncomfortable in anything more formal than corduroy slacks, open-collared dress shirts, and high-top sneakers, Skolnick often seemed every bit the dreamy academic who cares for little but the love of learning. Yet, Skolnick's unpretentiousness belies a molten ambition fired as a young man.

Raised in San Mateo, California, about halfway between Stanford University and San Francisco, Skolnick met many "famous scholars and doers" who regularly visited his parents' home. His father was a psychoanalyst and also held a part-time teaching position at Stanford. Among the elder Skolnick's friends from the university who visited often was the Nobel Prize-winning biologist Joshua Lederberg. Skolnick's mother was a "political activist, a feminist from the thirties," he says. "She was fighting for unionization of workers, fighting for rights of blacks in the fifties and the Japanese in California. She was a liberated woman before it was fashionable." Skolnick calls her simply a "do-gooder."

Like Mary-Claire King, Skolnick had a knack for mathematics and, also, he enjoyed computing. As a result of conversations with his father's academic friends, he decided to pursue a career in research. After graduating from the University of California, Berkeley, he sought an advanced degree there in demographics. But soon he grew restless. "I realized I already knew what I could learn from the program and that if I stayed it was just to get a Ph.D., not because I was truly fascinated," he says. "I had this revelation that I wanted to solve problems, that what I was really good at was making observations and trying to understand what they meant.

And I wanted to have an impact on individuals and not pursue some abstract scientific inquiry."

In the spring of 1969, after only six months at graduate school, a frustrated Skolnick sought advice from Lederberg. During their many talks over the years while Skolnick was growing up, the scientist had piqued the younger man's interest in genetics and human evolution. Sensing that Skolnick wanted to incorporate genetics into his studies, Lederberg introduced him to an eminent Italian colleague, Luca Cavalli-Sforza, who had recently joined the Stanford faculty on a part-time basis. After many hours of conversation, the Italian agreed to take on Skolnick as a student and offered him several research projects. "The one I jumped at, of course, changed my life," Skolnick says.

Back in his native Italy, Cavalli-Sforza was trying to construct the "family trees," or pedigrees, of the many generations of people who lived in an isolated mountainous region near the northern city of Parma, where the scientist had his principal academic appointment. By studying the pedigrees, Cavalli-Sforza hoped to identify subtle genetic changes that occur over time in a population that was fairly homogeneous, largely because it hadn't been infiltrated by outsiders carrying their own specific gene variants. Cavalli-Sforza suggested Skolnick help him out by using his computer and demographic skills to organize centuries of family histories culled from extensive church and village records.

By the spring of 1970—six months after he arrived in Italy—Skolnick was fluent in Italian, teaching demographics at the University of Parma, and was courting Cavalli-Sforza's librarian, Angela, an attractive and inquisitive young woman. Skolnick was rangy in those days, wore bell-bottomed jeans, had sideburns, long hair, and an unkempt, bushy mustache. To Angela, Mark looked like "something from another planet," she says. "As soon as he walked through the door I thought to myself, 'There's someone I want to marry.' " Skolnick arrived in Italy in September; by January they were engaged and married by June. "I was an exotic hippie-looking American Jew," says Skolnick, trying to explain the attraction. "She certainly was exotic to me."

Not long afterward Skolnick, quite by chance, heard about "Morman genealogies" for the first time. One day, a colleague introduced him to three emissaries from the Genealogical Society of Utah who had arrived complete with 125 cameras to microfilm every official family document—such as government census data, births, baptisms, marriages, and deaths—located in the Parma region going back to the early years of the Renaissance. The society is an offshoot of the Mormon Church that for years had set itself the colossal duty of tracing the origin of every family in the world, and then storing and arranging the mountain of data in a library-like research center in downtown Salt Lake City, the headquarters of the Church of Jesus Christ of Latter-Day Saints, as the Mormons are formally known.

The Mormons explained to Skolnick that similar teams dedicated to the church were spanning the globe, dropping into communities such as Parma, as part of a seemingly impossible task to collect for posterity the names of every person from recorded time. One reason for doing this is that the Mormons believe it is their duty to help guide ancestors to salvation, which can only be accomplished by knowing their precise position in a family genealogy. Mormons preach that everyone is sealed for eternity to their ancestry through marriage and their children. Thus, it is every Mormon's obligation to gather one's family history as far back in time as possible. The richer one's recorded genealogy, the greater one is blessed. The global genealogy project is, therefore, a deeply religious mission for those who embark on it. It also provides the world's best collection of documents for almost anyone—and most especially western Europeans—seeking to construct a pedigree of their ancestry, a practice the Mormon Church believes can spiritually enrich everyone no matter their religion.

By then, Skolnick was producing a computerized program that effectively linked the various individuals and families in the Parma communities whose names and relations had been painstakingly identified by Cavalli-Sforza and his research colleagues. If the Mormons were correct, the kind of family data Skolnick was working with in Parma was being replicated from regions all over the world and being deposited in one central site, Utah. To Skolnick, now hooked by Cavalli-Sforza on the idea of tracking gene variants as they flowed through specific populations, the Mormon catalogue of family data produced an intoxicating allure. "I realized that if things worked out my life's work was in Utah," Skolnick recalls. "Not studying Mormons, mind you, but studying Swiss, or French, or Belgians, maybe. I had no idea I might someday study cancer."

The opportunity came more quickly than he ever could have dreamed. Back in the United States, Cavalli-Sforza was invited, as senior scientists often are, to help judge research grant proposals at the NIH. At the grants review session, he was asked an intriguing question by another participant, Gordon Lark, a faculty member at the University of Utah. Lark told the Italian that in addition to the worldwide names project, Mormons living in Utah had amassed an extensive set of thousands of genealogies of the many people who were descendants of the original Mormon pioneers. Lark was vaguely interested in tapping the genealogies to study the incidence of cancer in the region but was uncertain exactly how to pursue such an undertaking. Soon afterward Cavalli-Sforza visited Salt Lake City and told Lark he had the perfect student to tackle the problem.

Indeed, while on a vacation in Hawaii, Skolnick finally had figured out how to build a computer program to handle the immense Parma family data. While sitting on the beach, Skolnick watched a fisherman tending his net. Its crocheted, weblike design inspired him to design software that could skim along the matings and offspring of an extensive family, following routes akin to the pattern of the fisherman's net. Skolnick persuaded Cavalli-Sforza that this linking program would give him the tool the Italian needed to track traits through the Parma families. Cavalli-Sforza figured Skolnick might help the folks at Utah with a similar solution.

In late 1973, Lark contacted Skolnick, who by then was concluding the Parman project. Lark and his colleagues hoped Skolnick, then 27 years old, would simply identify cancer-prone families using information from a Utah register of cancer cases required by law to be reported to the state health department by hospitals. "Right off, I realized I could use the [cancer] registry and genealogy to track down loads more families," Skolnick says. "But then what?" From Cavalli-Sforza, Skolnick had learned an important lesson. "His outlook was that you try to wrap your arms around an entire population group, not just a few families here and there, and only then can you begin to see otherwise hidden genetic variations," Skolnick says. "That's what he was trying to do in Parma and, I decided, that was what I wanted to do in Utah."

Universities can hire researchers and pay them to teach classes. But the University of Utah had no money to fund a population study of the type Skolnick proposed. Lark told him that if he was able to win an award from the National Institutes of Health, he would be given an academic position at the school. Skolnick had never before written a grant proposal, so he sought out several experienced hands, including the chairman of the genetics department he would join, Homer Warren. "I used to say that between us we had an M.D. and a Ph.D.," says Skolnick. "But Homer had them both."

Out of their work came what at the time seemed an audacious request, a $400,000 grant to be paid out over three years. In it, Skolnick and his university advisers proposed to make a single genealogy of all of Utah, linking the state's 1.5 million residents back, child to parent and husband to wife, to the original 5,000 pioneers who first crossed over the Rocky Mountains and into the Great Salt Lake valley fourteen decades before. Moreover, once the genealogy was organized, the proposal said the researcher planned then to look for dozens and dozens of cancer cluster families by linking the computerized genealogy to a computerized program of the state's cancer registry.

The grant proposal was titled, "Cancer Incidence and Risk Factors by Kinship." Says Skolnick: "My advisers warned me absolutely not to use the word 'genetics' anywhere in the title or the proposal. [That's] because, they said, everyone knew cancer had nothing to do with genetics."

Despite the researchers' attempt to veil their intentions, the

NIH grants review panel saw through it. "It was roundly trashed" by some of the judges, says one of the judges, P. Michael Conneally, a human genetics professor at Indiana University. "They thought it was a crazy idea." Conneally finally stood up and said, "Let's give it a closer look. You know Cavalli is Cavalli, and this fellow is a student of Cavalli's." If what Skolnick proposed was possible, Conneally said, it was worth funding.

The panel decided to send out an emissary to conduct a so-called site visit to assess Skolnick and his idea. But Skolnick was in no position to defend his request. He had planned to return to Italy to complete his doctoral thesis detailing his "fisherman's net" concept. The university told him if he wanted the project, he would have to stay and defend his grant request. By now he and Angela had a young son, whom he had to move from Italy to Salt Lake City. "We had a three-year-old and Angie couldn't speak much English," Skolnick recalls. "And Utah was about as foreign to me as it was to her." Moreover, Skolnick was in a race for time; the NIH panel had scheduled the site visit just six months away, "a frighteningly short time for me," he says.

Skolnick moved his family to Salt Lake City in early February 1974. Within days he was scanning documents at the cancer registry, eventually recording all the cases of male breast cancer and, also, lip cancer compiled over 15 years, believing that since both those conditions were rare they might well be triggered by inherited mutated genes residing in a few Mormon families. He then went over to the Utah Genealogical Society and pored through "family group sheets" for each individual cancer victim. The group sheets were the handwritten pedigrees assembled by Mormon families and placed in the society's research library for other family members to use. After weeks of searching through various nuclear families, Skolnick was able to make an astounding discovery. Many of the thirty-three cases of male breast cancer were related, through some close and some distant relatives, to one another. Working by hand, Skolnick had uncovered a family in which it appeared that male breast cancer might well be inherited.

In fact, the data were too crude to prove anything, but they did persuade the site visitors, who arrived in August; they decided that the cancer registry and genealogical data were deep and consistent, and that by tying the two together in a computerized program, Skolnick might uncover similar families with other disease clusterings. The panel approved the grant. Skolnick was awarded his Ph.D. in human genetics in January and then was hired by the University of Utah as an assistant professor. He received the first portion of his grant on February 1, 1975, fifteen months after his initial visit to Salt Lake City.

Soon Skolnick learned that the Utah vein he was mining was richer than he supposed. "From the very beginning, the Mormon leaders did two things for which geneticists will be forever thankful," says Skolnick. The church encouraged followers to have large families in order to build their religion quickly. That was one reason why polygamy was accepted. For a century, Mormon families typically produced eight children, and today the birth rate among many in Utah is twice the national average. In addition, the church began its practice of tracking ancestors. For the first few years in Utah, Skolnick did little else but build the computerized linking of the 200,000 family sheets from the genealogy and tens of thousands of cancer cases from the state registry. Punching all the data into Skolnick's "fishnet" linking program took sixteen "person years," the equivalent of four people working for four years.

Soon after his first visit to Utah, Skolnick sought the advice of Nick Petrakis, Mary-Claire King's boss at the University of California, San Francisco. "I thought what he was attempting out there was a great idea," Petrakis says. But he also cautioned the younger scientist that "he shouldn't expect any fast results."

Skolnick, of course, knew the cancer genetics effort was a longshot; conquering the disease might well involve the great bulk of his career. Still, soon after he arrived at Utah he met several physicians who told him they often ran across patients with family histories of other illnesses besides breast cancer. "We soon were getting tips about melanoma [a common type of skin cancer], colon cancer, and, of course, various precursers to heart disease, such as high blood pressure and heart attacks, that were present in some area families at a rate that suggested some inherited factor was involved," Skolnick says.

Because Mormons, with ample-sized families were so abundant throughout the state, it wasn't at all surprising that doctors heard stories from their patients about aunts, uncles, siblings, parents, grandparents, and children stricken with the same ailment that brought the patient to seek out medical care. Once he realized that area health providers already possessed information that would lead him to families where numerous relatives were being buffeted by disease, Skolnick began networking aggressively in the medical community.

One doctor he met early on was Randall Burt, a gastroenterologist at Salt Lake City's University Hospital, who specialized in treating cancers of the bowel. Burt told Skolnick that when he treated somebody for colon cancer, especially someone with more than four or five siblings and numerous cousins, aunts, and uncles, he invariably heard tell of kinfolk with similar medical profiles. Indeed, after a while Burt began persuading relatives of his cancer patients to come to his clinic for examinations, sometimes catching—and successfully treating—early stages of the disease that hadn't yet been diagnosed.

"It wasn't unusual for me to find a brother or sister with an early stage of undetected cancer, considering that sometimes my patients had eight, nine, and ten siblings—some had even more. And, of course, each of their parents came from large broods. So now you've got sometimes dozens of uncles and aunts, and, of course, even more of their offspring." Burt also found that many of these relatives lived nearby, or at least within a few hundred miles of Salt Lake City. It was, therefore, relatively simple to track down and contact patient relatives. And since "people out here are by nature open and friendly," says Burt, "it often wasn't that difficult to get people to come in for a clinic visit when I told them what I was looking for."

Says Burt: "A woman came to me because she had bleeding in her stool, a symptom something was wrong. As we discussed the problem—she had a small tumor growing in her colon—she offhandedly tells me her mother might have died from 'stomach cancer.' So I ask if anyone else in the family had died of cancer. 'Oh yes,' she says. 'My older sister had a problem, and I think one of my brothers, no maybe two of them, had something wrong. Come to think of it, cousin Sue, my mother's older sister's daughter, and maybe her brother, too, had some kind of [bowel] surgery." If inheritance were at work, Burt says, there was ample evidence of its effects out in the towns and farm villages lining the western slopes of the Wasatch mountain range.

Hearing these tales excited Skolnick, persuading him that by studying the Mormon families he could launch a genetic understanding of all common diseases, from diabetes to heart disease to arthritis, not just cancer. The bounty of relatives in the Mormon clans made it practically impossible to ignore that, at least in certain families, something biological was having an impact on those people sitting out on various branches of the family trees.

But Skolnick knew he'd need an early success. He worried that his four-year federal funding would be drained before he proved the genealogies' worth. "In science, sometimes you have to tackle what's possible to prove you are on firm theoretical ground," Skolnick says. "Dreams don't win you research grants, results do."

So, in 1976, Skolnick decided to drill into the genealogies to help solve a genetics mystery confounding two colleagues at the University of Utah. He was setting the scientific stage for a groundbreaking insight that would change the history of gene hunting.

The disease in question was hemochromatosis, a disorder in which an individual's cells and tissues becomes awash in excess iron because the body can't properly process iron consumed in the diet. People with an advanced form of the disease can be identified by their peculiar bronze-colored skin. More seriously, the iron overload eventually leads by middle age to cirrhosis of the liver, diabetes, and heart damage. For many years, doctors thought the condition was caused by alcoholism, too much iron in the diet, or maybe a vitamin deficiency. Then in the late 1930s, doctors who noticed the disease among siblings and their cousins decided it could be inherited.[1]

But an important debate then ensued. When scientists examined families where hemochromatosis existed, they couldn't determine whether it was due to a single dominant gene passed along from just one affected parent or the result of a recessive gene, in which case two genes, one from each parent, must be inherited. The issue wasn't simply a bit of scientific curiosity but had important medical implications. If the cause was a dominant gene, doctors could test iron levels of the young offspring of an affected parent. If the child had unusually high iron levels in the bloodstream, doctors could assume the child inherited the disease. The damaging effects of the disorder could be postponed, perhaps, by monitoring these young people's iron levels and periodically bleeding them when their iron levels got too high.

If, however, the disease was recessive, young people with mildly high iron levels might have inherited only one copy of the deleterious gene. Scientists suspected that people in that situation, who had two versions of the gene, one healthy and one that causes disease, tended to build up iron, but in a degree not considered dangerous. They wouldn't need to be monitored or bled because they would never develop the full-blown disease. If the disease were recessive, however, someone who carried two disease-causing genes, one from each parent, would also be at high risk of developing full-blown hemochromatosis.

Soon after Skolnick arrived in Utah, he learned of the hemochromatosis quandary from two university doctors, Corwin Edwards and George Cartwright, who had long been interested in the issue and knew of few cases among the Mormons. The Utah researchers faced the same dilemma as had Mary-Claire King. They needed some kind of easily measured "polymorphic" gene marker, some visible or detectable trait that was almost always inherited with the disease. If the trait were consistently linked to the disease in many people and families, it was possible that the gene for this trait and the one for hemochromatosis were being inherited together. If the two genes were thus "linked," tracking the marker trait from parent gene to child gene could serve as a proxy for tracking the inheritance of the hemochromatosis gene. Of course, such a marker must vary sufficiently in its forms that scientists can tell which form of the marker—and the closely linked gene—an affected child inherited from either the father or mother. Just about when Skolnick was hearing about the problem, researchers in France suggested a solution.

The French researchers were using a crude trial-and-error method of testing several potential traits with known diversity, eventually testing human leukocyte antigens, or HLA protein types. When the French team reported hints of an association between hemochromatosis in some families and certain HLA types, other scientists attempted to duplicate their findings. Soon came evidence that in a particular family, one type of HLA was consistently shared by all those who had excessive iron levels, while the same HLA proteins often were missing from relatives with normal iron levels. This suggested that certain HLA genes were tightly linked, located so physically close on the same chromosome to the unknown and hidden hemochromatosis gene that it was likely the two genes were being passed together from parent to child. If true, the particular HLA protein might serve as a surrogate for following the still-unidentified gene as it flitted through a family. And following the marker would answer the question of whether the disease was dominant, in which case an affected child needed only one copy of the gene from an affected parent, or recessive, in which case two copies, one from each parent, were required.

By now Skolnick had produced a computerized linking of the Mormon genealogies in Utah using his "fishnet" program. He decided that helping to settle the hemochromatosis question was a good test of the power of his program. Tracking down many more relatives was necessary if the scientists were to increase the statistical strength of the HLA linkage suspected by the French, and then to show whether the marker—and the linked gene—was inherited in a dominant or recessive fashion.

In a study that would serve as a model for later gene searches, the computerized genealogy database revealed 250 relatives of 10 previously identified hemochromatosis patients, including distant branches of the family with third and fourth cousins, some of whom were unaware of their kinfolk's disease or their own potential risk of having it. In one case the computer informed the researchers that there were 117 near and distant relatives of just one patient. The researchers then contacted the family members identified by the computer, advised them about the study by mail or phone, asked for their permission to participate in it, and then, when they agreed, sent out health aides who retrieved blood samples from them. The researchers tested each person's iron levels and typed their HLA protein variant.

Immediately, Skolnick and his colleagues realized an unexpected benefit. The blood tests uncovered fifty-five relatives of hemochromatosis patients who also had abnormally high iron levels. Seven of them had levels so high they were alerted, for the first time, that they had the disease. "To be honest, I was surprised and, of course, thrilled that the research identified so many people who had no idea they were at risk of the disease's consequences," Skolnick says.

Skolnick and a graduate student, Kerry Kravitz, then set up a computer program testing for linkage between those people with high iron levels and their HLA. As the results rolled in, it became increasingly clear that the hemochromatosis gene was recessive. The researchers identified twenty Mormons with high iron levels that could only have been due to inheriting two genes, one from each parent. Some of these people, of course, already knew they had the illness. But in eight instances, involving young adults, the disease had yet to progress far enough to be identified. Warned of their impending risk, the individuals were immediately bled to deplete their iron stores and, quite possibly, stave off the disease's ravaging effects.

"The predictive power [of the study] was startling," says Skolnick. "It convinced us that if we could use the genealogy to solve the genetic susceptibility of other, much more common illnesses, we had the possibility of making an enormous public health contribution."

Linking the disease gene to the HLA protein produced another result that eventually made the experiment especially important to human gene hunting but was of little practical importance at the time. Since the genes for the HLA complex of proteins were known to lie on chromosome 6, the researchers assumed the hemochromatosis gene probably resided somewhere on the same chromosome. Thus, the genealogy database helped the scientists to "map" the location of the disease-causing gene.

In early spring of 1978, at a small gathering of researchers in a mountain resort outside Salt Lake City, Skolnick got an opportunity to describe his group's gene-mapping coup to several leading figures in genetics. It was this fateful meeting that helped scientists devise a method for finding hundreds and hundreds of polymorphic DNA markers, the breakthrough that ushered in the new age of gene hunting.

Each April, several University of Utah faculty members held a week-long retreat at the Alta ski lodge about an hour north of the university in the Wasatch mountain range. There, graduate students and post-doctoral research trainees described their experiments. Several veteran scientists from other institutions also were invited to critique the research-in-progress. Skolnick's group was excited to present its hemochromatosis gene-mapping discovery to two rising stars in biology invited to the 1978 meeting, David Botstein, then at Massachusetts Institute of Technology, and Ronald Davis of Stanford University.

Botstein and Davis were among a handful of researchers perfecting a new set of laboratory tools, certain natural enzymes that were being used to study and manipulate DNA. Several years earlier, biochemists had figured out that bacteria regularly use certain enzymes in order to survive in hostile environments. If a particular bacterium has developed a mutant gene that makes it resistant to an assault from a powerful antibiotic, other nearby bacteria naturally seek to acquire the same protective version of the gene. The bacteria do this by using enzymes to snip apart a site in their DNA into which the protective gene is then inserted.

The biochemists soon learned that these so-called "restriction enzymes" slice DNA at predictable spots along the sequences of base-pair letters that constitute DNA. A restriction enzyme does this by scanning the DNA until it comes across a specific series of base-pair letters. For instance, whenever a particular bacterial enzyme comes across the series of base letters "G A G C T C," it always slices the DNA open exactly between the T and the final C. Botstein, Davis and a handful of other DNA pioneers were identifying these bacterial enzymes and using them in test tubes to slit open stretches of DNA in order to splice into them "foreign" bits of DNA or whole genes. Indeed, this new enzyme-based "gene-splicing" technology that was developed during the late 1970s opened the way for genetic engineering and helped give birth to the biotechnology industry.

During the meeting at Alta, Skolnick's researchers lamented, just as Mary-Claire King, that the dearth of polymorphic markers was the major obstacle to identifying human disease-causing genes. Botstein and Davis said the solution already existed. Without much prompting they both explained to Skolnick and others that it was very likely they could use the restriction enzymes to create a flood of useful polymorphic markers.

The two biologists explained that often when they spliced bacterial DNA with the enzymes, they uncovered, at certain specific spots, fragments of genetic material of varying base-pair sequences. The researchers suggested that these DNA variations likely existed in humans, too. Although Skolnick and his group weren't familiar with the technology being described, "we understood the implications," Skolnick says. "We were very excited, although we weren't really sure we believed them. Finding good markers had become this huge problem, and here these guys were saying they'd found the solution, right there, right before us."

As Botstein and Davis explained, the polymorphic (that is,

physically variable) DNA markers likely existed between the "cutting sites" of certain restriction enzymes. In human DNA, the biologists reckoned, the lengths of DNA base-pair letters that were snipped out by the enzymes would likely vary from individual to individual. Thus, one person's DNA spliced out at a specific site by a particular enzyme might, for example, yield the following base-pair sequence:

G A G T C T T T C G C A A T C T A G (Fragment One)

But a fragment removed from the exact same site in another individual might vary. Certain letters might be inserted or deleted from the DNA. Thus a second's person's DNA at the same site might look this this:

G A G T C T T T A T C A G C C G C A A T C T A G (Fragment Two)

In this example, the *underlined* letters are what makes the otherwise similar length of DNA variable, or polymorphic, from one person to another. Such variations later became known as "restriction-fragment-length polymorphisms," or RFLPs.

Skolnick immediately understood that, if these RFLPs were sprinkled liberally throughout the genome then it was possible that researchers could use them as "genetic landmarks" to hunt for the presence of a gene believed to cause an inherited disease in a family. For example: If a mother in a family had a disease, such as breast cancer, scientists would try to find some polymorphic bit of DNA, a marker, that could be used to trace the genetic material she passes on to her daughter. Every person carries two copies of each of the twenty-three chromosomes that contain all human genes, receiving one copy of each chromosome pair from each parent. Scientists then look for RFLP markers, selected randomly from chromosomes throughout the genome, until they find one useful in tracking genetic material as it is passed from parent to child. The marker would be useful because its length would vary in the mother and father and could be used to distinguish which one of a mother's two chromosomes were passed to the daughter, and which chromosome came from the father.

Thus, if a mother and a child both have an inherited disease and also share an identical form of an RFLP marker, scientists can begin to suggest that a gene responsible for the disease might be located on the inherited chromosome where the shared marker resides. But for scientists to say convincingly that an RFLP marker's site on a chromosome is associated with a disease, they would need to repeat this parent-child connection in a large number of large-sized families to create powerful statistical proof. If all diseased relatives in a particular family all carry the same version of a marker, and if healthy relatives consistently don't, and if this association is repeated in other families at the same marker site, only then can scientists argue that the marker's location and the disease are, in fact, statistically linked. The researchers can also propose that this marker-disease linkage may be due to the existence of a gene hidden somewhere nearby, in the same area of the chromosome where the marker is located.

Almost immediately, Davis, Botstein, and Skolnick realized they had to prove two things. The markers had to exist in humans, and they had to be located throughout the DNA. Soon afterward, the researchers recruited Ray White, a young molecular biologist at University of Massachusetts to find a human RFLP. In 1979, working with an associate, White found the first RFLP. Within a year, White landed a position at the University of Utah, where he set up a laboratory to find markers all over the genome, a map of landmarks for gene-hunting scientists.

Indeed, within a few years of the RFLP breakthrough, dozens of markers uncovered by White and others helped researchers discover genes underlying such inherited maladies as Huntington's disease, cystic fibrosis, and muscular dystrophy. Thanks to the RFLP technology and DNA polymorphisms, scientists searching out genes decided to launch a major effort to identify all 100,000 or so human genes, eventually creating the international Human Genome Project and the National Center for Human Genome Research run by the U.S federal government.

The discovery of RFLPs also made it possible for scientists such as Skolnick and Mary-Claire King to prove, for the first time, that common cancers could be linked to inherited genes. The existence of RFLP marker also caught the attention of a young and ambitious scientist in Baltimore who was struggling to understand the precise biochemical mechanism by which a cell turns cancerous.

9: Guardian of the Genome

F**ocus!**

The word is scribbled in crayon, graffiti-style in foot-high letters, on the white wall of a laboratory practically hidden behind locked doors in a renovated brick supermarket on the fringes of Johns Hopkins University medical center in Baltimore. The laboratory, just a maze of tiny rooms and offices, and hallways that tumble into one another, is where a small troupe of gene scientists led by Bert Vogelstein changed forever the direction of cancer research.

Beginning with a quiet trickle of reports in 1985, and then bursting in a cascade of findings from 1989 to 1994, the laboratory was by the reckoning of many one of the most prolific and, perhaps, the most influential of any in the cancer gene field. In 1993, and then again in 1994, *Science Watch*, a newsletter that tracks trends in basic research, anointed Bert Vogelstein the world's "hottest scientist" of the year.[1] Second in the newsletter's ranking in 1993, just behind Vogelstein, was the lab's codirector, and Vogelstein's protégé, Kenneth Kinzler.

By 1994, the 46-year-old Vogelstein was widely regarded as a leading candidate for a Nobel Prize.

More than 300 graduate students apply for the three to five research slots open every year at the lab. "This is one of the most exciting and extraordinary places to conduct research," says Todd Waldman. "There may be no place in molecular biology like it."

Much credit for the lab's success is due to the word that Vogelstein one day hastily scrawled in an arc around the room's thermostat, an unwitting metaphorical gesture, to be sure, for the intense scientific heat that it was meant to convey as well as generate. "I don't care what kind of crazy idea you want to pursue," Vogelstein tells new recruits. "But whatever you choose, make certain that's what you focus on. There's a million things that can distract you, there are probably ten, twenty, thirty ways to ask a question. Pick one, focus on it. Ignore everything else."

Vogelstein's own single-minded, fifteen-year focus on understanding the step-by-step genesis of cancer was impressive even among scientists absorbed by the cancer gene hunt. If indeed Vogelstein does win a Nobel prize, fellow researchers say he likely will be cited for a simple but persuasive notion about cancer progression that he relentlessly pushed his lab to uncover and then to extend in dozens of impressive, and often, groundbreaking experiments.

These experiments showed, beyond doubt, that cancer first arises in a single cell and then expands into its deadliest manifestations only after the cell and its progeny acquire and accumulate defects to a set of critical genes that govern the cell's life cycle. This is the cornerstone research underlying the "tumor suppressor" theory of cancer. And it came first from the labs of Bert Vogelstein. "Bert is the single most important basic researcher in cancer genetics," says Francis Collins. "Nobody else comes close."

By the early 1990s Vogelstein's multistep model of cancer genesis, so rational, elegant and provable, was embraced by nearly every major basic cancer research lab in the United States and aboard. It soon became the foundation for the most innovative research into the biological causes of cancer of the lung, breast, prostate, pancreas, and colon, the major cancer killers in the United States and elsewhere.

All this was extraordinarily satisfying to Vogelstein, a man of many passions and contradictions. In public he can be diffident

and guarded; small talk makes him seem awkward and unsophisti-
cated, and he concedes he is more comfortable in his dimly lit office than in celebrity's glare. But in private, among his family,
friends, and fellow scientists, Vogelstein displays a playful and bit-
ing wit, by turns clever, ribald, and sophomoric. His short, slender,
almost gaunt frame belies a tautly muscled body, broad shoulders,
and oak-thick arms honed by years of competitive tennis. His
reed-thin voice, punctuated with a cackling laughter, is unimpres-
sive, but whenever he opens his mouth inside his lab everyone else
is quiet. Anyone meeting Vogelstein for the first time, seeing his
thinning sandy-gray hair and unkempt beard, might think him an
offbeat poet or reclusive artist, not the powerful force behind one
of the world's most successful and innovative molecular biology
labs.

This presumption is easily reinforced by Vogelstein's measured,
low-key style. His typical conversation is sprinkled with self-
deprecating jokes. The only reference to his scientific success in
his office is a faded photocopy of an award from a pharmaceutical
company. Most available wall space is taken up with a poster of
Frank Robinson, the Hall of Fame outfielder who played for Vogelstein's beloved Baltimore Orioles in the late 1960s and early
1970s.

Indeed, Vogelstein and his lab members pride themselves on
being innovative, surprising and off-beat. Nothing exemplified
that more than Vogelstein's fanciful invention of a full-fledged
rock band in 1996 that, after many long, late-night practice sessions, eventually played to standing-room only crowds of Hopkins
colleagues in a grimy bar on the Baltimore waterfront. With Vogelstein at the keyboards (he is a trained classical pianist) and the
lab's co-director, Ken Kinzler, on drums, the band played old 60s
rock tunes and original blues music.

One of Vogelstein's most famous eccentricities is his "lab uni-
form." Every day, Vogelstein comes to the lab wearing the same
dark blue pants, white tennis shirt, white socks and black Reebok
sneakers. "I don't want to be bothered thinking about clothes,"
Vogelstein once explained. "I like to keep things simple," he says.

Indeed, until the summer of 1995, Vogelstein never used a computer; he didn't even have one in his office, relying instead on an IBM Selectric for writing grant applications, research papers, and lectures. ("I just don't have the time or interest in learning to use a computer," he said.)

Even his bland, vegetarian, and strictly kosher diet reflects a near-ascetic, uncluttered approach to life. While the graduate students and postdoctoral candidates in his lab swarm over pizzas, burgers, or pastries in their makeshift "cafeteria" just outside Vogelstein's office, their leader usually stands aside, leaning against the doorway, confining his lunchtime diet to a bagel, some bits of tofu and sliced tomatoes. If he gets hungry later, he reaches into the drawer of his desk where he keeps a jar of fruit jam and crackers.

And while other renowned scientists often accept invitations for talks and lectures, the perquisites of fame and power that often produce free trips to spas in Europe or resorts in the Rockies, Vogelstein is the rare senior scientist who seldom strays from his lab. He often arrives at 5:30 in the morning and stays until 6:30 P.M., leaving "early," he says, so he can be home for dinner with his wife and three school-age children. Although his lab staff is twenty years or more his junior, few can match his pace. During one particular grind, an intense race for a gene in the summer and fall of 1993, "some of us formed the early-bird club to see who could get here before Bert," says Nick Papadopolous, a postgraduate student. "None of us ever beat him through the door."

His most defining characteristic, however, is his competitive drive and an obsessive desire to win. "No one works like Bert," says Papadopolous. "I don't know anyone that does what he does." Following a two-year intensive assault on an inherited colon cancer gene that culminated in the spring of 1994, in which Vogelstein drove his lab mates relentlessly, Papadopolous is honest about his ability to keep up a "Vogelstein" pace for the rest of his career. "I don't know what I'll do when I leave the lab," he says. "This field is very, very competitive. I like to win, but I don't know if I can do what it takes to win in this field. I don't know that I can be like Bert."

Vogelstein's strength and dedication, some close friends say, go back to his religious upbringing. His grandfather was the rabbi at an old Orthodox synagogue in Baltimore, and even when Bert was a youngster, Friday nights and Saturdays were for religious services, solitude and study. Despite the frenetic pace of his lab, Vogelstein still never works on the Jewish Sabbath, putting the time aside for reading and contemplation. Both his sons had their Bar Mitzvahs at the sacred Wailing Wall in Jerusalem. "It meant more that way," Vogelstein says. "To my kids, to me and to my family. I couldn't see doing the kind of glitzy deals so many people do. That's not me."

Never one to travel far from home, Vogelstein went to college at the University of Pennsylvania in nearby Philadelphia, returned home to Baltimore for his medical degree, internship, and residency in pediatrics at Johns Hopkins, and then in 1976, he went just forty miles down the highway to Bethesda, Maryland, for postdoctoral training in molecular biology at the National Cancer Institute. By then, the young Vogelstein had settled on a career choice: while treating young cancer victims at Hopkins he decided he could have the greatest impact by studying the underlying cause of the disease.

"I could have become a pediatrician; it would have been satisfying at the time," Vogelstein says. "But I felt, through basic research, I'd touch a lot more kids." Moreover, at the time cancer research was hot. Only a few years earlier, the federal government had undertaken its famous "war on cancer," pumping unprecedented funding into basic research, much of which was gobbled up by new molecular biology labs springing up in academic institutions all across the country. In addition, "the cancer cell was a black box," Vogelstein says. "Nobody knew what was going on inside there. There was some evidence it was a result of genetic mutations. But that was a hypothesis. The proof didn't exist. Back then I can't say I knew how to get at the evidence. But I did know it was an area where it was possible to have an impact."

Vogelstein had begun to reason his way into a career-long strategy for shining a light into the mysterious darkness surrounding the cancer cell. In 1971, Vogelstein had married Ilene Cardin,

whose family was deeply involved in Baltimore's business, philan-
thropic, and political scene. So in 1978, seeking a chance to com-
bine his new research interest and stay close to his family,
Vogelstein applied for a junior position in a new cancer research
program being set up with federal seed money at Johns Hopkins
School of Medicine. "Baltimore is my home," he says. "I wanted to
be here."

It wasn't long after Vogelstein joined the cancer center labs that
"he started coming up with whole new ways of doing things," says
Donald Coffey, a researcher who was Vogelstein's supervisor. By
then, Vogelstein had rejected two of the three prevailing theories
about what caused cancer. Some scientists believed then that cancer was caused by a breakdown in the body's immune system, its
natural disease-fighting mechanism. Under this notion, scientists
argued, a glitch in the immune system might allow errant cells to
escape detection from "sentinel" cells in the bloodstream that seek
and destroy unwanted germs and other foreign invaders. A
healthy immune system, scientists believed, would perceive a
growing cancer cell as abnormal and cast it out as if it were an un-
welcome virus or bacteria. A defective immune system would be
blind to the cancer growth.

A second idea then current among basic scientists was that
most cancers were caused by mysterious viruses that somehow in-
tegrated themselves into a cell's nucleus. Once there, the theory
went, the viruses wrench control of the cell, forcing it to replicate
uninhibited into a tumor mass.

Neither of these explanations, even if true, described to the satisfaction of Vogelstein and others what happens to the cell's genes
to cause unwanted cell replication. Even if a virus drives cell
growth into high gear, or even if the immune system can't identify
or deactivate such accelerating cell division, something unknown
inside the cell's nucleus, which holds the DNA, is at the root of
both problems. "By 1980 or so, I was pretty convinced the key to
figuring out cancer was at the genetic level," Vogelstein says.

There was some evidence for this line of thinking. Researchers
found that survivors of the atomic bombing in World War II who

developed cancer late in life had contracted strange rearrangements of their DNA. In addition, there was a growing certainty that radiation caused cancer by damaging DNA. Some of the first medical technicians working with x-ray materials—who touched radiated particles with their hands—developed skin cancers; and women with tuberculosis administered high doses of radiation later developed breast cancer at a higher than expected rate.

"I thought these were all clues that the problem in cells was in genes," Vogelstein says. "But I could only prove that by actually finding the genes that were affected and showing that, in cancer, they were mutated."

One other idea was swirling about. Why, many wondered, did cancer usually take so long to emerge? Even people exposed to radiation didn't develop disease for a decade or more. And most cancers don't arise until late in life. Vogelstein suspected that was because a cancer cell had to collect several mutations. A defect to one gene might be dangerous, but not until the same cell sustained damage to other genes was it likely to spin out of control, he theorized.

Indeed, the suggestion that cancer involved several genetic steps had been put forward, but with only scant attention, by an obscure researcher in Texas. In the early 1970s, Alfred Knudsen, then at M. D. Anderson Cancer Center in Houston, suggested that a rare childhood eye cancer, retinoblastoma, was caused when the cells in the retina sustained two "hits" to a specific gene within the cells. He speculated that if a child was born with one of two copies of the defective gene (either the one inherited from the mother or the one inherited from the father) the child was at risk of disease. But if during early childhood, when the retina cells are dividing rapidly, a random error occurs in the production of the otherwise healthy copy of the gene, the double-damage to the genes allows the cell to become cancerous.[2]

This "two-hit" idea obsessed Knudsen. In 1976, he and others showed that children with the eye cancer had a loss of bits of their DNA on chromosome 13, suggesting that loss of a particular gene or its function was at the heart of the disease. Then in 1983, while

Vogelstein was searching for a way to study genetic machinery of a cancer cell, scientists elsewhere produced a landmark finding that dazzled the fledgling field of molecular genetics. Researchers working with Ray White in Utah, using his new RFLP markers, showed that a specific area of chromosome 13—likely a gene—was missing in children with the eye cancer. The scientists discovered this by showing that in these children an RFLP marker in the chromosome 13 region that should have been present was consistently absent from tumor tissue, suggesting that the marker resided in a part of the DNA that was deleted.

"It suggested that the gene in that site somehow was critical to sustaining a healthy cell, and when it was gone, the cell became cancerous," Vogelstein says. "Now that was the kind of proof I wanted to find in my cancers, too." The RFLP marker "was the critical" tool to finding "missing genes" in other cancers, Vogelstein says.

By 1983, Vogelstein had decided that if cancer was as Knudsen described, a multiple-step disease involving the progressive loss or damage to certain genes, then he needed to study human cells as they incurred these various genetic changes. At about the same time, an elderly physician at Johns Hopkins named Ben Baker had convinced a group of young researchers to tackle colon cancer. Baker gathered around him a team that called itself the Hopkins Bowel Tumor Working Group, and "he gave us some of the resources as well as the encouragement to take on colon cancer," Vogelstein recalls.

Back then "Bert didn't know a colon from a kidney," says one of the group's members, a young pathologist named Stanley Hamilton. "But he had this idea of looking for genes in various stages of cancer progression, and we told Bert that colon cancer might be the best disease available to undertake that kind of research." Indeed, within a few short weeks of becoming part of the working group, "I was convinced colon cancer held the key," Vogelstein says.

Vogelstein was attracted to colon cancer for several reasons. First off, "I soon learned from Ben that it was a major public health problem," he says, noting that about 130,000 Americans developed the disease each year. About 45,000 people in the United States annually succumb to the disease, often because the cancer isn't discovered until it reaches an advanced stage and has spread elsewhere in the body. From a research perspective, colon cancer was important because its stages of development were identifiable and, perhaps as important, accessible to researchers.

The first sign of a colon cancer is blood in the stool. Tipped off early enough to the presence of a problem, doctors can insert a catheter into the colon to see and then remove the disease's first precancerous stage, when it is merely a polyp. This mushroom-shaped tiny growth over the years is fertile ground for other molecular changes that can force the polyp to grow progressively more dangerous, eventually exploding into a fast-dividing carcinoma, which is the first sign of a tumor. Vogelstein figured that, as with the eye cancer, damage to genes might be what allows normal cells to expand into a polyp, and that an accumulation of additional genetic losses might transform a benign polyp into a cancerous form.

But studying the colon cells as they went through such changes "was something nobody else had even thought about doing," says Hamilton. "I don't think Bert was prepared for what we had to do next."

In order to make headway against colon cancer, Vogelstein needed samples of tissues removed from patients. Nobody had ever tried to analyze genetic alterations in human cancer cells directly removed from tumors. Previously, researchers had studied animal cells, or human cells cultured painstakingly over the years in test tubes. But Vogelstein was convinced that looking at genetic changes within human tissue was essential. He enlisted Hamilton, whose pathology lab regularly received and analyzed cell abnormalities in biopsies of tissue suspected as being cancerous.

"Whenever we heard that someone was being operated on for colon cancer, Bert and I rushed over to surgery and stationed ourselves directly outside the surgical suite," Hamilton says. The two men then took tissue samples and began immediately sifting through each one to find those cells that were cancerous. In order to do this Vogelstein's team had to invent a technique for "purifying" the tumor tissue.

Back at the labs, Vogelstein and Eric Fearon, a postdoctoral researcher, set to work looking for RFLP marker deletions in DNA extracted from the purified colon cancer cells. "What we were looking for was differences between healthy colon cells removed from the patients and cells removed from their tumors," Vogelstein says. But looking for marker changes throughout the genome was an impossible task. Scouring the medical literature they soon found a clue that they hoped would shorten the hunt. Scientists in France looking at colon cancer tissue under microscopes had found large sections of chromosomes 17 and 18 missing. So, one by one, Vogelstein and Fearon tested RFLPs up and down chromosome 17.

Finally, in 1987, they made an electrifying discovery. In 22 of 30 tumor samples they studied, markers in one particular spot in chromosome 17 were consistently missing. When they went back and studied polyps from the same patients they found no such chromosome 17 abnormalities. To the scientists this meant they were on the track of a gene that, when damaged or missing, transformed a polyp growth into a cancer. In a research paper published that October, the scientists said that "on the basis of the data presented here and elsewhere, one can begin to formulate a hypothesis for the development of colorectal [cancer]."[3] The formal language of scientific research papers doesn't allow a scientist to commmunicate emotions. But for Vogelstein's lab "the chromosome 17 discovery was a watershed," recalls Fearon. "It seems like everything came like a rush after that."

By 1988, Vogelstein's lab was running much like an assembly line. Vogelstein had as many as a dozen postdocs, graduate students, and technicians whizzing about the labs or poring over experiments at their lab benches. Every day, the scientists would examine another bit of chromosome 17 and 18 and 11 and 5—all spots where DNA losses in colon cancer tissue had been identified—trying to determine if, as they now fully expected, specific genes were involved in tumor growth. Standing over test tubes, the scientists examined bits of tumor samples still flowing from Hamilton's pathology department. They then digested the samples with special enzymes and extracted DNA. Then, once more using special molecular biology techniques, they fished out of the DNA those segments of the four chromosomes under study and examined the RFLP markers mixed in with the DNA.

By this time, "we were pretty convinced we were on the track of a tumor suppressor, a gene that kept a cell from becoming a tumor when it wasn't defective or damaged," says Ken Kinzler, who had joined the lab.

By the middle of 1988, Vogelstein's clutch of researchers had produced evidence that they were tracking specific genetic changes that occurred at identifiable stages as the cancer progressed from an adenoma, or benign polyp, until it reached a full-blown tumor. By examining 172 samples with dozens of RFLP markers, the lab found missing bits, or mutations, in chromosome 5 in about one-third of noncancerous adenomas, and they found mutations in a gene on chromosome 11 in about half the carcinomas. The labs found no evidence of deleted or missing DNA in chromosome 17 or 18 until they peered at samples removed from end-stage, lethal tumors. In those deadly forms of disease, the lab consistently found that bits of DNA were absent in these chromosomes.

To Vogelstein the data were clear. Certain specific and identifiable genetic changes sent normal colon cells spiraling, first, into a benign growth, and then, as the genetic changes mounted within a cell, into an out-of-control tumor. "Our results support a model in which accumulated alterations affecting at least one dominant-acting [gene] and several tumor-suppressor genes are responsible for the development of colorectal tumors," Vogelstein and his team announced to the world in a study published in September of 1988 with much fanfare in the prestigious *New England Journal of Medicine*.[4] In an essay solicited by the *Journal* for commentary on the implications of the new insight produced by the Vogelstein team, Peter Nowell, a cancer researcher in Philadelphia, wrote the following: "At the molecular level, an outline is beginning to emerge of the series of genetic events involved in a fully developed human cancer and of the underlying mechanisms."[5]

Of the report, Nowell says, "I had only heard vaguely of Bert before seeing his study. It certainly put him and his lab on the map. It seems like, after that, we heard something new from Bert almost every other month or so." Indeed, in his commentary Nowell wrote that it seemed clear from the Vogelstein research that some of the genes identified were part of a new class of tumor suppressors. But he threw down a challenge, too. "Unfortunately," he wrote ". . . none of these suppressor [sites] have been isolated thus far, and their locations and activities have largely been surmised."[6]

Indeed, Vogelstein felt with the release of the *Journal* data that he finally had confirmed his original idea: cancer was inspired by a specific set of genetic alterations. The lab had collared a rogues' gallery whose actions had turned a well-mannered human tissue into cancerous outlaws. Moreover, he had uncovered a novel scenario for human cancer development that "seemed to resemble the process of Darwinian evolution."[7] Each gene mutation seemed to free the cell from one of its normal growth controls, giving it a growth advantage over its neighbors, to replicate and divide more often than cells without the mutations. As the cell split itself over and over, it created frequent opportunities for additional mistakes, or mutations, to emerge during the production of new DNA required for each new cell offspring. With each new mutation, the cell lost another growth regulator and became more likely to outpace the growth of normal cells nearby, robbing them of circulating nutrients to feed its own tumor-massing needs. Thus, the cells with the most mutations overtook a neighborhood, by turns becoming a cancerous tumor mass. Because these replicating mistakes, or mutations, were rare events, Vogelstein proposed that it might take a cell years to accumulate the precise genetic miscreants that drove it to misbehave.

But that wasn't enough. Vogelstein also knew he had to find the cancer-causing gene itself. And he had to show how its loss of activity triggered cancer. Even before the *Journal* article was published, Vogelstein threw himself full throttle into the effort to track down the gene. The Human Genome Project had not yet been formed and only a handful of genes had yet to be discovered through the use of the RFLP marker technology. But Vogelstein felt he was as close as any research team to nailing the identity of a fabulously important gene. And now, his drive for success and achievement surfaced with a force that even Vogelstein found surprising.

"I know from our conversations that he felt he was on the verge of making a very big discovery that could really help people in a way that was difficult to discuss," says Martin Abeloff, director of Johns Hopkins Oncology Center.

Vogelstein started coming into the lab at dawn, turning on the lights and the heat, and starting experiments before anyone else arrived. In a quiet but forceful manner, he spurred his colleagues to keep up with him. "He was a ball of energy and ideas," says Fearon. "And he was sort of whipping us into this frenzy of work."

He knew once he published the marker's location he would have competition among others who sought to make their name by cloning the gene. To give himself an advantage, he began rummaging once again through scientific reports, searching for any hint of unexpected gene activity on chromosomes 17, 18, or 5 that was linked with cancers in reports by other research groups. It was then that Vogelstein came across a startling set of nine-year-old research reports that had languished in the backwaters of biology research. The library discovery he made on a December day in 1987 was to rank about as important as any he later made in the lab.

Back in 1979, biologists Arnold Levine at Princeton University, and David Lane at the International Cancer Research Fund in London[8] had independently, but simultaneously, sequestered a previously unknown chemical protein while studying how certain animal viruses cause tumors when injected in mice. The protein they found was attached to a part of the virus, suggesting to Levine and Lane that the chemical played a role in triggering cancer. In 1982, another group of scientists identified the gene that made the protein, and called it p53, which stood for "protein with a molecular weight of 53."

At the time, Levine and Lane didn't realize they had uncovered a mutant form of p53. Instead, they assumed that something about the normal version of the gene, when it was perhaps somehow activated by a virus and pushed into a high level of expression that it initiated cancer. "It was a funny mistake, looking back," says Levine. Indeed, it confused Vogelstein. Reading about the gene's discovery he was perplexed. It looked as if p53 was located smack in the middle of where his RFLP markers were pointing to a missing gene. But the earlier studies by Levine and Lane suggested that p53 was a gene that inspired cancers only when intact and unusually active, not when it was mutated, entirely missing, or damaged.

Throughout the summer of 1989, the Baltimore lab undertook a slow molecular crawl down the entire length of the short arm of chromosome 17, where they believed their culprit gene was hiding. At first, Vogelstein didn't believe the address where the markers had delivered him and his team. The spot was the home of p53. But he and his lab colleagues didn't believe p53 was their culprit. Finally, Sue Baker, a graduate student, spent several months sequencing all the nucleotides in the DNA they had located. After comparing the sequence she produced to the published sequences of p53, Baker discovered that the stretch of DNA found by the lab—and known to cause cancer—differed from the normal p53 in just a few of the gene's thousands of nucleotides bases.

"We just didn't believe it was possible that such a subtle change in the gene could unleash cancer," Vogelstein says. "It took us a long while" to believe p53 was the gene the lab had been tracking.

So Vogelstein had to conduct still one other experiment to prove that mutated p53 played a significant role in cancer. "We knew we'd cause a real ruckus with our claim unless we had the goods, the evidence we were right," he says. The lab then examined tumors from 56 patients. Cells extracted from healthy tissue of all the patients had two normal functioning versions of p53. But 75 percent of the patients' tumor cells had just one copy of p53, and that one copy was slightly mutated. The discovery was riveting, for it mirrored so closely the "two-hit" notion Knudsen had discovered years before regarding the childhood eye cancer. When two copies of the gene were present, the cell was tranquil. But when one copy of p53 was missing, and the other slightly damaged, the cell grew into a tumor.

When the experiment was reported in April 1989, "it changed cancer research forever," says Stephen Friend, then a young, up-and-coming molecular biologist at Harvard University and Massachusetts General Hospital in Boston. "Nothing's been the same in cancer since. It certainly changed my life and the work I do." Indeed, as a result of the paper, Friend wondered whether p53 was guilty of causing the string of serial cancers of the breast and other tissues that were afflicting a group of families uncovered by scientists at the nearby Dana Farber Cancer Institute. Sure enough, in 1990 Friend found that in these families, parents were passing along a mutated version of p53 in their egg or sperm cells. With one copy of p53 already screwed up, once the second copy experienced a slight, but nonetheless lethal, mutation, Friend discovered, a cancer emerged.

"The researchers had been looking for a gene underlying the [breast] cancer syndrome for years," says Friend. "Then along comes Bert and turns the world upside down."

Scientists at the National Cancer Institute and elsewhere began examining tumors from cancers of the lung, prostate, bladder, and brain. "We were stunned, simply stunned," says Curt Harris, who ran a large lab at the government cancer center. "We found that p53 was mutated in about half of all the cancers we looked at. You know, for years we'd come to believe that cancer wasn't one disease, but was many diseases with many causes. Ten years ago if you'd have told me half of all cancers would have one mutant gene in common I'd just never believe it."

By 1992, dozens of labs all over the world were conducting experiments to understand how p53 did its dirty work. It turned out that p53 is a "Jekyll and Hyde" molecule, says Harris of the NCI. "It sometimes can be very good and sometimes very bad" for the cell.

That's because the protein p53 makes was found to be a so-called transcription factor, an enzyme that latches onto other bits of DNA and directs other genes into action. In a series of discoveries, scientists, including those in Vogelstein's lab, found that p53 acts like an emergency brake. As a cell goes through a series of steps before it divides, p53 is able to determine whether certain genes in the cell-division process are functioning correctly. If these genes aren't, p53 is able to alight onto the DNA of the cell and literally freeze the division process until the malfunctioning genes can be repaired. Indeed, if the cell-dividing genes are found to be in a state of complete disrepair, p53 initiates a series of steps that forces the cell to commit suicide.

Of course, if p53 itself is damaged, none of this monitoring can take place. Then cells normally under the watchful eye of p53 divide incessantly, or in a manner consistent with tumor growth. That's why, in 1992, David Lane, who codiscovered p53 in 1979, dubbed it "The Guardian of the Genome," for its critical role in protecting and monitoring genes in the all-important cell-dividing process.

Subsequent research found that a faulty set of p53s was involved in fifty-two different kinds of cancer: in 50 percent of cancers of the brain, in 35 to 60 percent of bladder cancers, in 30 to 40 percent of breast cancers, in 90 percent of cancers of the cervix, 30 to 60 percent of ovarian cancers, and, of course, 40 to 70 percent of colon cancers. In one especially enlightening set of experiments, researchers found that ultraviolet rays from the sun can cause cancerlike damage to skin cells by damaging p53 in their cells. And the same lab also found that an ingredient in cigarette smoke likely causes cancerous damage to lung tissue by impairing p53. An entire school of research soon emerged, testing known carcinogens to see if they act by causing p53 defects.

Meanwhile, other researchers began exploring the possibility of harnessing p53 to treat cancer. In test tube studies, scientists began inserting healthy versions of p53 into cancer cells. "What they found was truly incredible," says Ken Kinzler, by 1992 codirector of Vogelstein's lab. "In each case, replacing a mutated p53 with a [normal] one stopped tumor growth cold." This experiment drew the attention of the pharmaceutical industry. "It told us that if we could somehow replace or mimic the action of p53 we would have a remarkably powerful anticancer medicine," says Frank Mc-Cormick, a biochemist at Onyx Pharmaceuticals in San Francisco.

At the same time, the Baltimore lab crew continued to hunt for the other hidden genes Vogelstein and others suspected were involved in the colon cancer cascade. If, as they began to believe, a healthy colon cell is transformed into a tumor through the gradual accumulation of several mutant genes, then Vogelstein was convinced that a defective p53 gene was only the first of these genes to be uncovered. "In colon cancer, at least," Vogelstein says, "p53 appears to be one of the last of several genes that must be altered to cause a cell to become cancerous."

Indeed, from their cancer cell analysis Vogelstein soon gathered evidence that DNA markers were often missing in chromosomes 5 and 18, as well as chromosome 17, where p53 was found, suggesting to the researchers the location of the additional genes involved in colon cancer. In 1990, Eric Fearon found the chromosome 18 gene. And in 1991, Ray White in Utah and Vogelstein collaborated to find a gene on chromosome 5 that causes an inherited disease in which dozens of polyps sprout in the colons of people born with the defective form of the gene. This inherited disease, called familial adenomatous polyposis, or FAP, is relatively rare, accounting for about 5 percent of all colon cancer cases. But the gene was an especially important discovery because Vogelstein and others found that even in the more typical cases of colon cancer that are not inherited, the tumor cells contain defective forms of p53, the chromosome 18 gene, and the FAP gene.

By 1992, "we were beginning to get the clear picture of the molecular steps a cell goes through—the accumulating of specific defective genes—in order to become cancerous," Vogelstein says. To him and others, cancer was a black box no more, although mysteries still abounded. And, in their effort to solve them, Vogelstein and his lab soon found themselves in a controversial and competitive race that would reveal a type of gene never-before seen in humans, and a form of cancer genesis never imagined.

10: Ishmael's Tale

His name was Ishmael. He lived along the northeast coast of Newfoundland in an isolated fishing village that was hammered in winter by nasty blasts from the north Atlantic Ocean. And he had a story to tell, a kind of haunting family tragedy that by the early 1990s was providing pivotal clues for solving the mysteries of cancer.

Luckily for Bert Vogelstein and his gene-hunting crew, among those people to whom Ishmael chronicled his sad story was a curious and sympathetic geneticist named Jane Green, working in the Newfoundland capital of St. John's. Green was so captivated by Ishmael's narrative that she set off by car hundreds of miles into the thinly populated Canadian frontier to investigate it. Several years later, her research into Ishmael's family history, cluttered as it was with lives tortured and cut short by colon cancer, became a major turning point in the quest for a type of colon cancer gene that had dodged detection for almost a century. Indeed, for years biologists and cancer doctors had argued, often ferociously, over whether the common type of colon cancer in Ishmael's family was truly caused by the inheritance of a gene at all.

"Jane Green's family was the key for us," Vogelstein says. "Without Jane's family we would not have made our discovery. There is no doubt or ambiguity about that. Many doctors, especially doctors who saw evidence of the gene in families they treated, believed the [colon cancer] gene existed. But in science, of course, believing isn't enough. You have to find the gene, you have to identify its structure, you have to prove it is implicated, that people with a mutant form of the gene get cancer, and those with working versions don't."

Indeed, when the Vogelstein lab and a competing group finally identified the new gene and several of its chemical relatives in a flurry of eye-popping discoveries between late 1992 and early 1994, the mechanism by which the genes triggered cancer awed scientists. The discovery of the gene revealed that cancer sometimes was caused by a flaw in an ancient, but hitherto unknown, biological process that was bequeathed to humans through eons of evolution, from microbial ancestors that first began scrambling about the planet's surface hundreds of millions of years ago. That the gene could be traced to the beginnings of animal life suggests to scientists that it is critical to human health.

"When a gene is conserved that long you've got to believe it's important, very, very important," says Richard Kolodner, a Harvard biologist. Kolodner was one of the first researchers to isolate the gene in microbes, such as yeast and bacteria, years before anyone had evidence that it played a role in a devastating human cancer. And for months before the gene was discovered, he was to play a leading role as a mystery researcher shadowing Vogelstein in his effort to capture one of modern-day biology's greatest scientific trophies.

The gene discovery showed that human DNA contains a "quality control" department. DNA's quality control guardians are made up of certain human proteins that are generated in cells by specific genes. These proteins have the ability to scan for and to correct mistakes in the multitude of letters and words in the genetic messages that control cell behavior. This quality control mechanism, a sort of biological spell-checker, assures that the three billion letters that make up a person's complement of DNA are reproduced with precise fidelity each time a cell divides.

Scientists long suspected human DNA contained some type of

genetic "proofreader" monitoring the minute yet massive job of DNA reproduction. But without identification of the genes or the proteins they express there was no way to know for certain that they existed, nor was there any way even to guess exactly how these quality assurance proteins carried out such a colossal assignment. Indeed, most cancer scientists didn't know the biological process of correcting DNA spelling errors existed. The few claims, made in arcane scientific journals by microbiologists studying germs, that such editing proteins might play a role in cancer genesis were largely ignored or unknown to human biologists.

"Every time we wrote a grant we said studying . . . repair genes in yeast would likely produce invaluable insights into cancer," says Harvard's Kolodner. "We weren't broadcasting the claim, and, of course, we didn't have any way to prove we were right. It's understandable that for years our work went unnoticed by the big cancer guys." Although Kolodner carried out his research in laboratories in the famous Dana Farber Cancer Institute in Boston, and much of his research was underwritten by pharmaceutical companies and government agencies who believed his work might inspire new revelations into cancer, he was unknown to Vogelstein, Francis Collins, or any of the scientists with a reputation for hunting for cancer genes.

But when the cancer scientists stumbled upon the existence of the DNA mistake-detection and repair system in humans it changed the way they thought. "We didn't set out to find [repair] genes; we didn't know they existed," says Ken Kinzler, noting that the lab was mostly interested in working out the precise mechanism of the p53, FAP, and chromosome 18 genes. "But that's what we found. In retrospect the involvement [of repair genes] in cancer makes perfect sense. You couldn't have predicted it, but maybe we should have."

The gene hunters found that when repair genes are themselves defective, they are unable to carry out their critical quality assurance jobs. Normal production mistakes that cells regularly make when replicating DNA are allowed to pass unnoticed, and random errors of the type one might expect in any gigantic copying job occur. In other words, a flawed repair gene can allow flaws to go unchecked in many other genes, some of which may have the crucial job, for example, of regulating the normal growth cycle of a cell, telling it when to divide, when to stop dividing. When genes that oversee the life cycle of a cell are damaged, and cell growth is allowed to continue without restraint, dangerous masses of cells can accumulate over time, and can eventually grow into a deadly cancerous tumor.

Proof that defects to the repair genes are a major underlying cause of colon cancer culminated in "the most extraordinary set of coincidences in a short period of time that I've ever seen in scientific research," Vogelstein says.

The importance of the gene discoveries did not go unnoticed by the scientific or lay press. In the spring of 1993, when the latest type of colon cancer gene's existence was first confirmed, Eric Lander, a leading geneticist at Massachusetts Institute of Technology, hailed it in a *Science* magazine news article as the "most exciting development in human genetics in the past year."[1] When, six months later, the gene's exact chemical identity was unmasked, Stephen Thibodeau, a Mayo clinic researcher, told *The New York Times*'s Natalie Angier, "This is a terribly, terribly exciting discovery." In the same article, which the *Times* heralded on its front page, Francis Collins said, "This is a grand story, a wonderful discovery."[2]

But while scientists lauded the find, others worried that its use as a tool to predict future disease was certain to raise troubling implications. Inheriting a defective version of the gene not only raised the risk of colon cancer, it also put women at a higher-than-usual risk of uterine and ovarian cancer. While uterine cancer can be cured if caught early, there is no way of identifying ovarian cancer until after it strikes, often not until after it has begun to spread. And once ovarian cancer spreads, it can be especially deadly. Thus, doctors worried that women in families with the colon cancer gene might feel pressured to have their ovaries removed in order to eliminate the cancer threat. But scientists had no way of telling women who carried a defective form of the gene whether

they also carried a risk of ovarian cancer, or when it might develop.

"The [colon cancer] gene discovery is freighted with the same
promise and peril of so many gene discoveries these days," says
Neil Holtzman, a Johns Hopkins University professor who is an
expert in ethical issues related to genetics. "Until there are out-
right cures, and there are none right now to combat the colon cancer gene, we are stuck in a troubling time. We can identify people
at risk before their cancer arises, but we can do damn little for
them. I don't know any doctor who feels comfortable with that. I
don't know any family who feels reassured by such an advance."

The pursuit of the repair gene was launched in earnest in 1989
when Bert Vogelstein attended an international cancer gene
meeting in Japan where he met Jukka-Pekka Mecklin, a geneticist
from Finland. Mecklin told Vogelstein he and his colleagues at the
University of Helsinki were tracking a form of inherited colon
cancer that differed dramatically from familial adenomatous polyposis [FAP], the type of colon cancer that flourishes amidst an explosive growth of polyps in the colon. Vogelstein was well aware of
FAP and by 1989 was already helping Ray White of Utah search
for its genetic cause. Instead, Mecklin said he and his research associate, Albert de la Chapelle, had strong evidence of a different
type of familial colon cancer that was unrelated to any exceptional
amount of polyp growth. "The cancers are there, the pedigrees tell
us they are inherited, but there are no inherited polyp problems [as
in FAP]," Mecklin told Vogelstein.

Vogelstein was intrigued but doubtful. He had heard vaguely of
doctors in the United States, of Henry Lynch in Nebraska, most especially, who also claimed that they had uncovered a family-related
colon cancer with no evidence of polyp growth. But in 1989, Lynch's
claims weren't quite dismissed, but they weren't given much cre-
dence either. That was so despite the fact that, as with his claims
about familial breast cancer, Lynch had produced a mountain of reports describing colon cancer bunching up in certain families, which
he labeled the "Lynch syndrome." Its more scientific name is "HN-
PCC," for human nonpolyposis colon cancer.

By 1987, Mecklin reported a study of 122 patients who had developed colon cancer by age 41.[3] These cancers struck families in a
pattern that suggested a dominant defective gene was at work,
putting members of these families at risk. But Mecklin's attempt to
screen people at risk was one of only a few reports in the medical
literature by the time he and Vogelstein met in Tokyo. At the
meeting, Mecklin was deeply impressed by Vogelstein's use of
RFLPs and other markers to track down genes, and he was convinced his families would provide Vogelstein the raw material to
make a gene hunt possible.

Finding a gene underlying the Finnish families would be "no
straightforward or simple" assignment, Vogelstein knew. "There's
no stigmata, no unusual or easily identifiable physical defect that
allows you to distinguish who has the gene and who doesn't," Vogelstein commented at the time. Moreover, since colon cancer is
so common, it had been difficult to determine whether clustering
of the cancer in a family, absent the polyps, was simply due to coincidence, not an inherited gene. When looking for a gene by sifting through the DNA of several families, "if you mix in a few
pedigrees who have the disease simply by chance, not by inheritance, you are going to foul up the entire effort," Vogelstein says.
"That happened to us a lot."

Still, Vogelstein was impressed by the large size of the pedigrees
Mecklin collected. Once back in Baltimore he phoned de la
Chapelle, setting up a collaboration. Finland's isolation and its
largely homogeneous population had become an especially rich
environment to discover disease-causing genes. Most of Finland's
descendants can be traced to prehistoric farmers who settled in the
southern part of the country, arriving from the European continent about 2,000 years ago. Indeed, most of the country's present-
day population of five million residents are descendants of about
250,000 "founders" who lived in Finland around the year 1500.[4]
Geneticists call such an occurrence an evolutionary "bottleneck,"
a period in which a specific population in the past can be identified as the principal source for a much larger genetically related
group of people living today. De la Chapelle exploited this bottle-

neck as no other geneticist before him. He already had more than a dozen disease-causing genes, each of which likely first arose in one of the founders as a spontaneous mutation between the years 1500 and 1700, and which was then passed along in certain spreading families, over the many years becoming the source of the many inherited diseases in modern-day Finns.

Within a few years, working with Mecklin and others, de la Chapelle had collected about forty colon cancer families, many of whom had ancestors the researchers could track back several hundred years. They did this by identifying a colon cancer patient who had one or more close relatives with the disease, and then tracing back the patient's ancestries from church parish registries, which in Finland are considered "official" documents, and which contain remarkably accurate records of an individual's birth and death, often even providing a detailed accounting of the precise nature of a particular ancestor's demise.

Thus, in one genealogical study the Finnish researchers carried out during their colon cancer hunt, they found that in a group of eighteen families, eleven were identified as descending through thirteen generations from a handful of ancestors who had all lived in a small rural region located in south-central Finland. Over time, it became apparent to the scientists that all eleven pedigrees actually shared the same tiny gene mutation, strongly suggesting that the colon cancer rampant among these families was probably derived from one founding father or mother who was born with a gene defect he or she passed along to numerous offspring in succeeding generations.[5]

This fact alone shows the immense power of gene-hunting science to reveal the past as well as help predict the future. "Interestingly, evidence of a single gene mutation shared by the eleven different families with colon cancer was great proof that these families, previously unknown to themselves, were all distantly related to one another," de la Chapelle says. "It's a fascinating outcome from the disease-gene research."

The speed with which colon cancer families were collected and documented in Finland produced the strongest evidence yet that colon cancer unrelated to the presence of numerous polyps was, indeed, a disease that could be commonly inherited. By the time de la Chapelle and his group had embarked on a colon cancer pedigree search, a small but growing group of molecular biologists were showing that it was possible to find hidden genes by scouring DNA removed from members of highly informative families. By 1992, the labs of Vogelstein and de la Chapelle had tested samples from several Finnish and American families, first testing for mutations in p53 and the newly discovered FAP gene. "I thought for certain one of those genes was involved," says Vogelstein, "but they weren't." Absent any further clues, the next step to finding a gene meant scouring the entire genome, from top to bottom, using whatever available markers and laboratory techniques were being developed. "It was going to be one heck of an effort" to find the genes, if they existed, remembers Ken Kinzler.

It was then that a fortuitous event occurred during a casual conversation on a lawn at an annual genetics meeting in the seaside resort town of Bar Harbor, Maine. Every year, in late July, the genetics department of Johns Hopkins University and Jackson Laboratory, a mouse genetics lab down the road from Bar Harbor, sponsor a two-week intensive genetics course for scientists, genetics counselors, and doctors interested in incorporating genetic science into their research or clinical practice. The sessions are presided over by Victor McKusick, famous as the father of genetic mapping at Johns Hopkins. Many of the lectures are presented by the university's faculty members, who are encouraged to come with their families.

"People hear all sorts of things and meet all sorts of people from fields of inquiry they wouldn't normally come in contact with," says David Sidransky, a Johns Hopkins researcher. "It's a stimulating meeting in all ways."

The 1991 meeting in late July came just weeks after Vogelstein and White had found the FAP gene, the polyp-causing gene, and its discovery dominated many conversations. Indeed, there was a session on colon cancer that included a discussion of the p53 and FAP genes. Several researchers from Vogelstein's lab attended, and so did Jane Green, then a 46-year-old geneticist at St. John's University Medical Center in Newfoundland.

"Following the [colon cancer] session I was out on the lawn, enjoying the sun and drinking coffee," remembers Green. "I just happened to be standing near Laura Hedrick and Dan Levy, two members of the Vogelstein lab. They were excited, talking with great enthusiasm about having just found the FAP gene, and talking about how they had found so many of the genes in the colon cancer pathway. Then one of them, I don't remember who, said, 'Now all we have to do is find the Lynch gene.' That really got my attention."

The Vogelstein lab researchers "lamented to a group of us standing nearby that the research" into the Lynch syndrome cancer-causing gene was stalled because of a "lack of really good, informative families," she says. The researchers explained that linkage studies were hampered because many of the affected members of Lynch syndrome families were deceased, and their DNA was unavailable.

"You could have a great family on paper," says Green. "But if the affected members are dead, ascertaining the exact cause of death and collecting and studying the DNA can be very difficult, maybe even impossible." As a result, too many of the families collected up until then in Finland and the U.S. had critical gaps that spoiled efforts by the molecular biologists to conduct DNA analysis in the labs.

Green, however, by this time had spent the entire past year investigating Ishmael's story, collecting information about his family, as well as another large family from a neighboring coastal region. "By 1991 I had spoken to everyone in both families, even some who had left Newfoundland," says Green. In all, she had contacted about 150 people. One family was composed of the descendants of five brothers who all had died in their late twenties. Some of the men's widows had moved to Nova Scotia and taken their children with them, many of whom were just old enough to begin developing cancer, too.

Standing on the green lawn, sipping coffee, Green briefly described the two families to Hedrick and Levy. So interested were they, the two researchers pressed Green to crudely draw out the pedigrees on a bit of paper. Hedrick and Levy couldn't believe their good fortune. "Sometimes science is a matter of luck," says Hedrick. "What Jane was telling us was simply too good to be true."

By that evening, Green asked colleagues back in St. John's to fax her a precise version of the family pedigrees, which Hedrick immediately dispatched by fax to Vogelstein in Baltimore. "I'd seen a lot of families by then whose promise fell apart under intense study," Vogelstein says. "I was interested, but I wasn't jumping up and down." Nonetheless, Vogelstein proposed to Green that she join his gene-hunting crew. And within a few weeks he telephoned Green several times, asking if she could collect DNA samples for laboratory analysis. Vogelstein knew it wasn't a simple request. But despite being located in a far-off site not known for genetics research, Green was a savvy investigator who already had collaborated in a cancer gene discovery.

By 1978, Green had begun running a clinic at St. John's University hospital that identified people at risk of developing inherited eye diseases such as retinitis pigmentosa, which causes severe vision impairment. Then in 1982, in the course of an ordinary referral from a doctor in a distant town, Green was thrust into the arena of cancer genetics.

Like the Utah pioneers and those in Finland, the people in Newfoundland had a tradition of large families. "When I arrived, it wasn't uncommon to meet families with 10 or 15 children," Green says. "As a result, there existed these regional genetic clusters. If a family shared a genetic disease, it was easy to observe. You'd see it in offspring, and in siblings, in their families, in cousins, and in all sorts of related kinfolk who all lived within a very short distance of one another. For a geneticist, these cultural characteristics produced an unbelievably rich living laboratory."

In 1982, the doctor's referral introduced Green to one of these gene clusters, several patients in one family with Von Hippel-Lindau disease, an unusual inherited disorder caused by a dominant gene passed along from one parent that almost always causes tumors of the eye, brain, and nervous system in a gene carrier. Green soon decided to track down the patients' entire family, believing that she could save lives by encouraging family members with an affected parent to undergo screening tests. Her hope was simply to catch undetected cancers in people unaware of their inherited risk. Within a few years, Green and her colleagues at the genetics clinic had identified forty-four relatives in Newfoundland with early 50-50 risk of inheriting the disease, and she was gaining the reputation on the giant island as someone to seek out on issues related to heredity and health. That's how she came to meet Ishmael Mast[6] in the late autumn of 1990.

One day a surgeon at the university medical center where Green worked introduced her to a middle-aged man—Ishmael—who was dying, following his second serious bout with colon cancer. The surgeon sought out Green after Ishmael complained that there was "too much cancer" in his family. Within a few weeks of hearing him lay out details of his family, Green was on her way to visit several members of the Mast family who lived near one another in a small fishing town in Newfoundland's northeast coast.

"Ishmael died shortly after I was introduced to him," Green recalls. But before that he put down on paper every bit of information he could find about his family, which he could trace back, in part, to six generations. "Ishmael had been a schoolteacher and he understood recordkeeping and where to find important documents," Green says. Among his papers, for instance, was a note, dated 1840, from the minister of the town's church, describing a visit to a man who had died of a disorder involving severe bleeding of the bowel. The man was Ishmael's great-great uncle, and, according to the note, he was 36 years old when he died. "Even in the harsh life these folks lived back then," Green says, "death from a bowel disorder at 36 was odd. It sounded very familiar to Ishmael, and much like the death and illness he was seeing in his family."

Despite the isolated location of Ishmael's home, and the 400-

mile drive from St. John's, Green visited there numerous times
during the next year. Soon she realized the area's regional hospital
held evidence of colon cancer in not one, but in two distinct families, each of which mostly lived in and around their own communities that were about 150 miles apart but were actually separated
by about 300 miles of rugged roads. Much of the relevant information she picked up was by interviewing Ishmael's relatives or members of the other nearby family, and then verifying what they told
her by examining medical records from local health providers or
by scouring documents in the archives of small medical clinics and
hospitals. In one instance, she was told about a man who sounded
as if he might have died of colon cancer decades ago, but for which
she had no documentation. Finally, one day while rummaging
through boxes of papers in one small hospital's storage room, she
stumbled across the man's name on an index card in a card file dating back from the early 1900s. Sure enough, he had died of symptoms that sounded like colon cancer.

All this family documentation was needed in order to verify
that the pattern of cancer in Ishmael's family was consistent with
the impact of a dominant gene. "I couldn't go around predicting
people were at risk, or making claims that they suffered from an inherited disease, without the pedigree evidence," which meant
tracking down information about as many of the people in Ishmael's six-generation family as was possible.

By the time Green met up with the Vogelstein lab researchers
in Maine in July of 1991, she had uncovered twelve people living
with cancer in one family and another fourteen with cancer in the
other. Within weeks of receiving the fax of the two pedigrees, Vogelstein was on the phone to Green, asking her to collect blood
and tumor samples from the affected individuals and blood from
some unaffected relatives. "I'd done that kind of work before," says
Green, "but it was going to be quite an effort, especially since the
two families were difficult to reach."

Moreover, interacting with the people in the family was time-
consuming and emotionally grueling. "It's not like working with
lab specimens," says Green. "These are people, and I began to feel

very close to and concerned about them. Now that they understood they had an inherited disease and something might be done
about detecting it early many people became very interested in the
work and what it meant to them. I couldn't just call them up and
tell them I wanted them to send me blood samples and get me
cancer biopsies; I would have to spend an hour or more on the
phone or in their living rooms, educating them about genetics,
about the disease, explaining what their risk might be, or what the
risk might be to their children. They were very cooperative, and I
felt this tremendous responsibility to help them deal with this information. I think it is impossible to imagine what kind of devastating news this was for some people."

By the spring of 1992, the Vogelstein and de la Chapelle labs
had received enough blood and tumor samples to begin analyzing
them, looking for places in the cells' genome where those with
cancer had a shared DNA marker, while those without cancer
didn't. "We tried without success the genes known to cause inherited cancer—FAP, p53—and then we moved on, screening hundreds of markers at random from chromosomes all throughout the
genome," Vogelstein says. "It's backbreaking work. There's no fast
or easy way to do it. You come into the lab in the morning and stay
until late at night."

In the Baltimore lab "war room," the tiny closetlike space that
was alternately a conference room, library, and makeshift cafeteria, Vogelstein told his small army of postdocs and graduate students, already worn out from other gene hunts, to gear up for
another assault. "Bert called it 'Mission Impossible,' " says Nick
Papadopoulos. From the war room, the lab researchers sped back
to the windowless maze of laboratories to conduct their experiments. "The atmosphere was crazy, very intense," says Papadopoulos. "I wasn't doing anything else. I'd work twenty hour days, go
home, sleep a few hours and come back."

But despite their efforts, neither lab was able to unlock the
gene's secret hiding place. Vogelstein kept telling Green that her
two pedigrees (along with another he had secured from a New
Zealand researcher) were impressive, but none of the families he

had from Finland, Newfoundland, or New Zealand appeared large enough on their own to propel the research forward.

Green suspected she had a solution. Soon after she began investigating the two families, Green believed several clues suggested that the two pedigrees were really two branches of one family tree. The first tip-off was that the pattern of cancers in the families was strikingly similar, fitting consistently into a pattern already identified by Henry Lynch. Both families also traced their origins to England. And although each family congregated around one of two remote villages largely isolated from one another by the more than 300 miles of rough, inland terrain, it was well known that descendants of the original English settlers in the easternmost village had at one time migrated westward by boat into the area where the second village was located.[7] Still, none of the families Green met shared a common surname, nor could anyone recollect a shared ancestor.

Eventually she found one when a family member produced a Bible containing a marriage certificate that clearly identified that in 1882, two ancestors of the two families had indeed married. "They were all one family, I just knew it," Green says, recalling her exultation. She later confirmed the family connection in government archives.

Green immediately called Vogelstein. "I asked her if she was certain, absolutely certain," says Vogelstein. "It made a great deal of difference" because it significantly increased the power of the studies. "When I told the lab, there was great excitement," Vogelstein says. "It came at a good time, because we were getting pretty down about ever finding the gene."

Indeed, even with the large, combined family, Vogelstein and de la Chapelle were getting frustrated. They had tested over 250 markers of various types, from labs from all over the world, and none had pointed the way to the existence of a defective gene shared by the affected members of the families. In October of 1992, a month after Green connected the large family, de la Chapelle visited Vogelstein and went over the data. "We were pretty disheartened," Vogelstein says. "We weren't certain what to try next. To be honest, I still was skeptical as to whether we were looking at an inherited disease. I had my doubts, especially since colon cancer is so common."

But, by coincidence, at the same time de la Chapelle and Vogelstein pondered the future of the research in Baltimore, Green was in San Francisco at the annual American Society for Human Genetics meeting. While there she received a message from her clinic that a 27-year-old member of the extended family was just diagnosed with cancer. She immediately called Vogelstein with the news.

"I can tell you we were about to give up," Vogelstein says. "But a 27-year-old just doesn't get colon cancer. That is very unusual. It had to be due to an inherited, defective gene. I told Jane we needed blood samples and tumor tissue from him."

On December 20, in a snowstorm, the young man's father drove 50 miles to a local hospital to deliver the samples. Green in turn mailed them overnight to Vogelstein. The Baltimore and Helsinki labs then began a new push, using markers that contained so-called double- and triple-repeats. Throughout the genome there are bits of DNA in which two or three of the A, T, C, G nucleotide base pair letters in a row are repeated in sequence over and over, such as ACTACTACTACTACT. These repeats serve as good markers, because in some families they are inherited in altered form, in which the sequences are repeated more or less often than in other families. The repeats are like fingerprints; rarely do two families share the exact same number of repeats. But because every human has the repeats, a particular version carried by someone with cancer, but not by those unaffected, would be a powerful marker for locating a gene.

In the early spring de la Chapelle called Vogelstein's lab to say that a triple repeat marker residing on chromosome 2, the 345th marker tested overall, produced a strong association. The researchers took the repeat and tested other families. "After two weeks it was clear," Vogelstein says. "We'd found the gene's location. We'd found the proof we needed. Lynch syndrome was an inherited disease."

In Omaha, Lynch was stunned. "I was excited, yes," says Lynch.

"But for me, it meant that in all these families I had helped collect, we now could do testing, we could tell these people for certain who was at risk and who wasn't. That's what it had meant for me all along."

His son Patrick, who is also a cancer specialist, at M. D. Anderson Hospital in Houston, was similarly excited. "What it means is that finally we'll be able to sink our teeth into a problem that we have been flailing at for years," he said at the time. "Prior to this the only way people knew they had inherited the problem was when they developed the cancer. And even then, we didn't have definitive evidence a gene was causing the problem. With this discovery, now we do."

Vogelstein and de la Chapelle were delighted. "You know, the labs did a great job, truly heroic," says Vogelstein. "But the truth is we can't do anything here unless we have good family material. Jane and the others deserve as much credit as we get. Maybe more."

On May 7, *Science* magazine carried two reports detailing the mapping of the colon cancer gene's approximate location.[8,9] And once more the press and the science establishment lauded the biologist from Baltimore for making still one more extraordinary find. "Bert was becoming the toast of the town," says Stephen Friend, the gene-hunting scientist from Massachusetts General Hospital and Harvard.

Especially important, the researchers generated headlines when they claimed that by their reckoning as many as 1 in 200 people may carry a defective version of the gene they had mapped. "That would certainly make this [gene] one of the most common causes of inherited disease—if not the most common," de la Chapelle told *Science* magazine in a new article that appeared the same day.[10]

A four-year tortuous search had ended. But Vogelstein knew another lay ahead. "Mapping a gene is important, especially for early diagnosis," Vogelstein says. "This disease is totally treatable if caught early. But if we are going to make real headway against the disease, we need the gene, we need to know how it works, if ever we are going to prevent it, or treat it effectively."

In April, before the discovery was published, Vogelstein called his crew together again in the war room and gave them marching orders. Ever competitive, Vogelstein "told us he wanted us to find the gene," says Ramon Parsons, one of the lab postdocs. "He didn't want any excuses."

But little did Vogelstein know that the next month, immediately after *Science* published the discovery, another research group, working without notice or fame, was about to give Bert Vogelstein the run of his life.

11: Clone by Phone

News of the colon cancer gene mapping reached Richard Kolodner the day it was published in *Science* magazine.[1] Sitting in his office on the eighth floor of the famous Dana Farber Cancer Institute in downtown Boston, Kolodner read the new research with heart-thumping excitement.

"They've got my gene," he said. "They've got my goddamn gene."

Indeed, they had. Without knowing it, Vogelstein and de la Chapelle had stumbled into the fringes of biology research, a largely mysterious and enigmatic place where Kolodner and a few other microbiologists had been doing pioneering science without much notice for the previous twenty years. Within hours of reading about the new genetic discoveries, the 42-year-old Kolodner, a self-described "nobody" whose small lab was tucked away in the recesses of the giant cancer center, decided to join the international hunt for the colon cancer gene's chemical identity. The moment he made that decision he kicked off a nine-month scientific dash the likes of which no one had seen before in the ferociously competitive field of cancer gene hunting. Voegelstein and de la Chapelle didn't know about Kolodner, nor did they even know he was pounding on their heels until five months later, by which time, it turned out, he was already a few strides ahead. A "lab rat" at one of the world's preeminent cancer research hospitals, Kolodner finally had a chance to show that the science he and his lab mates had been doing, unseen by the public and most of the scientific world, was in fact about as important as any basic research being conducted in the field of cancer, genetics, and human health. "You know I often would come to work and wonder, 'What am I doing at a cancer center?' " Kolodner says in his slow but confidently direct manner of speaking. "I loved the work, the exploration, solving experimental problems. But would it benefit cancer research? I didn't really know."

But now, Kolodner and a long-time colleague from the University of Vermont named Richard Fishel realized they were entering a very big-time, high-stakes contest. Physically fit—Kolodner is an accomplished skier with the stamina to stand at a lab bench poring over experiments for hours at a time—the Dana Farber scientist felt he had spent his entire career in training for a scientific sprint he knew he'd have to run in the next few weeks and months. Kolodner had watched colleagues at Dana Farber and at Harvard, where he held a position as professor, turn a single, major human gene discovery into a career-transforming event.

Divorced and childless, Kolodner's life was neat, organized and stripped bare of any unnecessary baggage. The lab was his life. In the warm months, he walked the mile or so from his home to the lab, often arriving by 7:30 A.M. in shorts, sandals, and T-shirt, and frequently not leaving until twelve or thirteen hours later. His constant companions were his sly wit and his dog, an English setter called Ligase, named after an enzyme that sews together bits of DNA, a critical tool in the gene-splicing experiments Kolodner conducted.

For years, Kolodner had to scour for funds to keep his small lab alive. Despite making important basic research contributions in the workings of DNA, and despite being widely lauded by his peers, every year Kolodner had to fight for grants from the NIH, from pharmaceutical companies, and from wherever else he could. To succeed in finding the colon cancer gene, Kolodner would have to become an instant expert in human genetics. That was no small feat. But anyone who might have thought to bet against Kolodner didn't fully appreciate his previous accomplishments, his talent, or his desire.

Despite lack of recognition as a big-name biologist, Kolodner was famous—at least among a small coterie of scientists—as one of the world's premier experts in yeast and bacterial gene biology, and as a master lab technician in the rarefied world of molecular biology. He was as creative a laboratory scientist as Vogelstein or any of the brilliant young people who had been tutored in the Hopkins lab.

Three years earlier, Kolodner and another researcher in his lab had discovered a gene in baker's yeast that did the "most extraordinary thing," he says. The gene, he called MSH, produced a protein in yeast cells that was a kind of virtuoso of housekeeping, the kind of little-known and obscure protein—a behind-the-scenes kind of player—that keeps DNA from fouling itself up. The gene's protein was a yeast version of a protein also found in bacteria in 1983[2] that could repair "mismatches" in DNA base-pair alignment. The gene was one of the "quality-assurance" genes that made certain DNA faithfully reproduced itself each time a cell divided.

Kolodner's lab actually identified the yeast version of the bacterial genes in 1989, but the field in which he labored was so collegial, "I didn't need to rush it into print, or even publish it with great notice," Kolodner says. "We presented the work, we gave credit where it was due and expected the same." He didn't even submit the work for publication until early 1992, and even then the obscure journal to which he sent it took nine months to get it into print.[3]

In the research, Kolodner describes the isolation of the protein that repairs DNA replication mistakes. When a cell divides, each of its two progeny takes one of the two strands of DNA from the original cell. Each new cell's nucleus then makes available the chemicals from which a complementary strand of DNA is produced. Each strand of DNA is composed of a string of nucleotides, the chemical letters A, C, T, and G, that run in a precise sequence. Sometimes the nucleotide or letter sequence of the new strand is produced out of order, so that the base-pairs are mismatched. Such a mistake often leads to the production of a faulty gene, and, in turn, a defective protein produced by that gene.

A mismatch in the nucleotide letters of a gene for, say, the protein insulin, if uncorrected would lead to a mutation that would cause pancreatic cells to manufacture poor-quality insulin or even no insulin at all which would result in diabetes.

To prevent such problems, nature, in its evolutionary wisdom, has produced biochemical spell-checkers, proteins that slide up and down DNA scanning for mismatches and repairing mistakes when found. The mismatch repair system was first identified in the 1970s by microbiologists studying DNA replication in bacteria. Kolodner entered this field of inquiry, first as a graduate student at the University of California at Irvine and later as a post-doc at Harvard, where he began to hone an expertise in biochemistry studying the DNA of the E. coli bacteria.

After arriving at Harvard in 1975 he spent many years studying DNA recombination, another emerging area of gene science. Recombination takes place when genes or bits of genes are reshuffled during cell division. Soon Kolodner was studying exactly how DNA molecules break apart and rejoin, and how they correct errors that are made during this process.

In 1989, Kolodner found the mismatch repair genes in yeast, MSH1 and MSH2. Kolodner's discovery was hailed among his peers because of the technical prowess and exhausting persistence that it entailed.

Says Kolodner: "I do love to do experiments. It gives me great joy."

Finding the mismatch repair genes in yeast provided strong evidence that the spell-checking system likely existed in higher organisms, perhaps also humans. By the spring of 1993, Kolodner and his colleague Rick Fishel showed off their technical skill by fishing out the homolog, or similar version, of the bacterial repair gene in mice, suggesting that all mammals had such a mistake-repair system.

But what riveted Kolodner and Fishel when they read Vogel-stein's mapping for the HNPCC gene was a passage in the *Science* report in which the researchers said they had come across a strange phenomenon they had never before seen in DNA taken from cancer tumors. As it turned out, Kolodner and others in the field had seen the event before, many times before, in yeast and bacterial cells where the mismatch genes were defective.

Vogelstein and his colleagues had tracked down the HNPCC gene to chromosome 2 by use of DNA markers that had a series of repeating letters. In one set of markers, the letters CA, for the nu-cleotide units cysteine and adenine, were repeated numerous times, so that a section of the DNA near to the gene went like so: CACACACACACACACA. Because this repeating sequence can vary from person to person without causing harm, it serves as a useful genetic marker. Thus if a mother and her child both have 10 repeats at the same site in their DNA, it is likely that the child in-herited that chromosome from the mother, not the father. By tracking the passage of the CA repeats, the researchers were able to identify the existence and approximate location of the HNPCC gene.

What startled and perplexed Vogelstein and de la Chapelle was that when they checked the tumors from people who had inher-ited the HNPCC gene, they found disruptions or changes in the CA repeats at sites throughout the DNA in those tumors. Indeed, when they looked for CA repeats in other cancer cells they found such "genetic instability" scattered all over the genome, some-thing they'd never before noticed.

Indeed, in the same issue of *Science* in which the HNPCC gene mapping is reported, Stephen Thibodeau, a Mayo Clinic scientist, coincidentally reported seeing this same genetic instability in the DNA of 13 of 90 colon cancers he examined at random, cancers from people not believed to be part of any colon cancer family. "That was a very interesting finding," Vogelstein says. "Because it suggested that whatever was wrong in these families might also be at the root of some colon cancers that occur sporadically, cancers that are not caused by an inherited gene."

Ultimately, Vogelstein and others hoped that uncovering a gene involved in familial cancer would also reveal elusive pathways involved in all cancers.

But, more immediately, the genetic instability seen in the genomes of people with HNPCC implied something else. It meant, says Thibodeau, that some unknown phenomenon was causing a disruption in normal DNA replication throughout the genome. Perhaps, the researchers suggested in the *Science* papers, the faulty HNPCC gene was the culprit, somehow causing these widespread DNA reproduction problems.

That suggestion is precisely what tipped off Kolodner and Fishel. "Mutant mismatch repair genes cause that kind of problem in yeast," Kolodner says. "I was certain what Bert had found was the human version of the mismatch repair genes identified in bacteria and yeast. What I had to do next was prove it."

Actually, it was quite possible he already had the human gene in his hands. Six months earlier, Kolodner and Fishel had begun looking for a human version of the mismatch repair genes, and by May they had a large fragment of one of three different repair genes that scientists had found in bacteria and yeast. "The description of the [DNA instability] was a revelation to Richard and me," Fishel says. "It meant there was a one in three chance we already had their gene."

Within days, Kolodner and Fishel, along with five other colleagues they recruited, initiated the hunt. Kolodner started coming to the lab before dawn and staying until almost midnight. He stood for so many hours at a time at his lab that his back ached. He rarely socialized, and cut his jogging by half. And beyond his small group, he told no one what he was doing. By July, Kolodner and Fishel had mapped the gene to its location in the human genome; it sat in chromosome 2, perhaps near the spot where Vogelstein and de la Chapelle had mapped their HNPCC gene.

Back in Baltimore, Vogelstein was relentless. He knew nothing of Kolodner or of mismatch genes, an uncharacteristic intellectual oversight he later came to regret. But he had mapped genes before and knew that taking the next step—actually cloning the gene it-self—was the prize. Trapping a gene was not something to be done in a relaxing manner. In the *Science* report the researchers put po-tential competitors onto the scent of the gene by identifying its general location on chromosome 2. Given that, Vogelstein as-sumed that other top-notch gene-hunting labs, probably some he didn't even know about, were going to try to use the powerful cloning techniques available to molecular biologists to claim the gene for themselves.

Vogelstein loved the competition, for its own sake and because it propelled the science forward. Vogelstein conceded in private conversations that being the guy in the lead, being the cancer gene hunter everyone revered, satisfied a deep hunger. In high school it was the tennis courts, in his professional life it was the court of public distinction; finding the gene was so much more than that, he said.

"The gene is going to be found soon," Vogelstein told colleagues in May of 1993. For that alone he felt a supreme satisfaction for himself, for Kinzler, for de la Chapelle, and his lab warriors. But beyond that he knew he was helping to change the world. Right away, hundreds of people would know whether they had the gene and had a risk of having cancer. For those in families with HN-PCC who had the curiosity and courage to ask, the answer was now within their grasp. And the Baltimore and Helsinki labs had made that happen. There was no doubting now that colon cancer would be conquered someday because of this discovery.

Vogelstein handed off the job of finding the gene to five post-docs in his lab. If they did the day-to-day grunt work, he was their captain, inspiring them, driving them, and, quite literally, scaring them. Each morning, the postdocs would meet with Kinzler and Vogelstein and map out their assignments.

When the Baltimore Orioles opened their new stadium at Camden Yards that year, Vogelstein decided to buy four season tickets in the upper deck behind home plate. After a tough day in the labs during that summer in 1993, one or more of the postdocs would rush over to the ballyard, grab a beer and a crab sandwich, and try, for at least a few hours, to care more about Cal Ripken than colorectal cancer.

But Vogelstein knew that winning the race to clone the HN-PCC gene was not going to be easy, even with his resources. That's because the genetic region in which they had to scour for HNPCC was large, about 20 million base-pairs long, roughly one percent of the entire length of human DNA. If the entire genome were compared to the earth, then the Baltimore lab had narrowed their search to California. But since the typical gene is about 30,000 base-pairs long, the gene itself was likely hiding on a small stretch of roadway, perhaps a few miles along Malibu Beach, or somewhere between Golden Gate Bridge and downtown San Francisco. Where to start?

Vogelstein asked several other labs for help. At the NIH, Jeffrey Trent had been searching through chromosome 5 for other genes using novel techniques he had recently developed. He was asked to chop it up in bits and produce fragments the Hopkins people could investigate in detail. A group in Denver that had studied chromosome 2 was asked to generate and supply markers in the region. They asked for and received help from genetics researchers at Los Alamos, New Mexico, and in France, as well as from Henry Lynch in Nebraska and his son in Houston.

By 1992, "the research world was filled with so many great molecular biology labs studying the genome," Kinzler says. "It was not like just a few of us were doing this. Everything can move so much faster when you can reach out to people specializing in certain areas. Bert helps because he is such a voracious reader of research, and because people talk to him and he talks to them all the time. You can't be a loner and get things accomplished, at least not quickly."

Within a few months, the lab had developed or received about two dozen markers in the region. But in order to be useful, the scientists needed to study DNA from affected family members who had recombination, a shuffling of genetic material, in that region. Only by tracing a marker as it jumped around the DNA, moving as DNA was recombined, could the scientists shrink the size of the search area. In order to find useful recombinations, the Baltimore lab realized they needed DNA from many more affected people than they already had.

"We decided to do some grave-robbing," Vogelstein says. He sent out the word to Henry Lynch, to de la Chapelle, to Jane Green in Newfoundland, to the New Zealand collaborators and to the genetic counselors in Baltimore. "We told them to go back to the medical centers where patients had been treated and try to find tumor samples from the deceased members of the families," Vogelstein says. At Hopkins, the researchers Stan Hamilton and Gloria Petersen were recruited to pore through the pathology department's huge basement storeroom that was a repository for bits of cancerous tumors preserved in wax dating back to 1880. Within just a few weeks, the scientists had amassed DNA from 50 additional patients. "For us, at that stage, it turned out to be absolutely key," Vogelstein says.

By mid-September, they had closed to within one million base-pairs of the gene. It was time to look at candidate genes, those genes already known to be lying within the identified area, to see if any of them were the delinquent HNPCC gene. But one by one, the few known genes fell out of the ever-shrinking region under study. Only one gene remained as an option. And it had come to the Baltimore labs only a few weeks earlier as a real surprise.

One morning, a few weeks after the *Science* papers were published in May, Vogelstein received a phone call, "out of the blue," from Paul Modrich, a Duke University scientist who had purified the protein made by mismatch repair genes in *E. coli* bacteria back in 1982. Modrich told Vogelstein that he had seen genetic instability of the CA repeats and similar strange expansions and contractions of other repeated letters in the DNA of bacterial cells in which mismatch repair genes were mutated or missing.

"Oh, sure, mutated mismatch repair genes," Vogelstein replied. "Sure, we knew that." But, Vogelstein later admitted, "I didn't have any idea what he was talking about. I'd vaguely recalled reading something about mismatch repair, but, really, I'd never heard about the research or the genes."

Modrich suggested that the Baltimore researchers send him some tumor samples and blood for analyzing DNA from patients with HNPCC as part of a collaboration effect. Modrich had developed a biochemical technique for testing for the existence of repair mechanism. "He called back a few days later," Vogelstein recalls, "and said that just as he suspected, the DNA from these patients' tumors was totally devoid of any mismatch repair activity." This was the first hard evidence that repair genes were altered in a tumor.

Indeed, it turns out, as Vogelstein suspected, the *Science* report caught the attention of many researchers, but not who he expected. Several other biochemists working in microbes and in human cancers had the same idea as Kolodner, Fishel, and Modrich, that the instability reported in the tumors of HNPCC patients was related to damage to the mismatch repair system. In June, Manuel Perucho, a researcher in San Diego, published a report noting that he, too, had previously seen an unexplained variation on the repeats of certain DNA letters in cancer tumors. In his report he suggested that the problems were likely caused by some underlying malfunction in the cells' DNA replication machinery.[4] Perucho went so far as to suggest that he had uncovered "a new mechanism" behind cancer, speculating that it might be the result of mutations to genes involved in repairing DNA replication errors.

Then in late August, Vogelstein received another phone call, this time from Tom Petes, a biochemist at the University of North Carolina. Petes said his lab was soon going to publish experiments showing precisely what Modrich and Kolodner had suspected but never demonstrated: that mutations to the yeast repair genes did, in fact, produce DNA instability that was exactly of the type the researchers reported seeing in the HNPCC patients. In September, when Petes's paper[5] was published, Tom Kunkel, another member of the small mismatch repair community of scientists, suggested in an accompanying article that damage to human versions of the mismatch repair genes was the likely cause for certain inherited diseases, such as colon cancer.[6]

"That did it for us," said Nick Papadopolous, one of the postdocs in charge of finding the HNPCC gene in the Baltimore lab. "By the middle of September we were pretty convinced we were looking for a human homolog of the *E. coli* genes."

Indeed, it was in the middle of September that Vogelstein finally heard about Kolodner. The Baltimore researchers were struggling to determine whether a human version of the yeast genes lay in the tiny area of chromosome 2 in which they now believed HNPCC was hidden. Finally, after several of the postdocs looked up past research in the field, they uncovered Kolodner's 1992 paper describing his discovery of the yeast genes. Using a method outlined in the paper, the Baltimore team found their gene.

"Without Richard's paper we were sunk," says Vogelstein. "I'd never heard of Richard before that, though, and I still had no idea he had any interest in HNPCC genes. To me, he was a yeast geneticist. Why would he be interested in colon cancer?"

Still, the next and crucial step was to show that the gene they now had in hand was consistently mutated in people with the familial cancer, and not in the members of the families who were healthy. To do this they had to find the mutation itself. Finding a mutation requires skill and cleverness. The gene was more than 40,000 base-pair letters long, and a mutation could be as small as one letter shifted out of order.

Kolodner was struggling with his own problems. Unlike the Baltimore crew, Kolodner and Fishel had had the gene in hand since the late spring. But he knew, too, that he had to show that it was chromosome 2 that mutated consistently in patients. DNA from patient samples, however, was the one thing Kolodner didn't have, nor did he know how to get it.

"Everywhere I turned, I realized that anyone with really good HNPCC families was already working with Bert," Kolodner says, noting it was the only time he felt truly disheartened. Says Fishel, "Bert had 'em all locked up."

Kolodner turned to cancer epidemiologist Judy Garber for help. Garber had spent years collecting and counseling families with genetic illnesses, especially those related to cancer. Reaching out to her contacts in the United States, in Sweden, England, and

Switzerland, to cancer specialists around the world, Garber was able to collect ten small families. Scouring their DNA samples, Kolodner became even more despondent. He could find no evidence of any damaged genes in chromosome 2 of these families. What next? By late October, Garber had procured the DNA samples of two more families. This time, when Kolodner tested the mismatch repair gene he found what he believed was strong, but not conclusive, evidence of mutations.

Kolodner wanted to be certain, and hoped that Garber would find him still more families. But, like the Baltimore researchers, Kolodner was now hearing footsteps. "I was worried that we'd be scooped," he says. "We decided we'd better go ahead and publish what we had."

On November 8, Kolodner and Fishel submitted a paper to the journal *Cell* announcing the discovery: a human version of the mutS bacterial gene, which he called hMSH1, was the cause for HNPCC. The paper, however, lacked any reference to a definitive mutation, so the journal's editors asked Kolodner to include what evidence they had turned up. The next week, after returning the revised paper to *Cell*'s editors, Kolodner received an unexpected phone call. It was Bert Vogelstein on the line, all the way from Israel.

Vogelstein says he had heard a few days earlier from Paul Modrich that Kolodner was looking for human versions of mutS and so he now called to learn the status of Kolodner's hunt. Kolodner stunned him with the news that he already had cloned the gene, that he already had found mutations of it in cancer cells, and that he already had submitted his research to the journal *Cell*, which he expected to publish the work in December. "That," says Vogelstein, "was a real downer!"

Vogelstein suggested that since it appeared that both teams had pretty much nailed the gene at about the same time, they agree to publish together. Earlier that year, Vogelstein had made a similar agreement with another research team that had uncovered the cancer-causing traits of another gene called p21, using different techniques but at just about the same time. "The suggestion that we coordinate publication wasn't that unusual," Vogelstein says, noting that the field had become very competitive and many labs were producing important results at about the same time. Nonetheless, Kolodner was concerned about sharing the limelight with the well-known Baltimore lab, and he told Vogelstein he would think about it.

Vogelstein was beside himself. His group had worked terribly hard; they had produced outstanding DNA work in a short period of time. He was proud of them and worried how they would react. He called from Israel and notified Kinzler and the rest what he had learned from Kolodner.

"When I called they were very excited and happy and very up about telling me their progress," Vogelstein says. "Like *that*, the call went from incredible excitement to incredible disappointment."

Vogelstein and Kolodner exchanged several more calls, but ultimately Kolodner told Vogelstein he wanted to continue to go it alone. "Listen, we both had come to this discovery from different angles and gotten there at about the same time," Vogelstein says. "I thought publishing the research together would be a very powerful statement. I really thought it would be fair. I give Richard all the credit in the world. I wasn't trying to take anything away from him. I didn't want to take anything away from my people either."

Vogelstein told his researchers to rush to finish what they had. In the next two weeks, in a marathon of work, they found five mutations in the human version of the mutS gene, which they agreed to call MSH1, the name the gene was given by Kolodner. In one instance, the lab looked at DNA from a woman who had colon cancer at age 42 and endometrial cancer at 44. Following weeks of analysis, the lab found that in her gene just one of the thousands of nucleotide units had been replaced by another. When the researchers went back and analyzed DNA from twenty-one relatives with disease, they found that each had the exact same letter substitution.

Importantly, says Vogelstein, "the substitution occurred at a site in the human gene that was identical to a site known to mutate in the bacterial and yeast genes. In other words, it was a mistake in a very important area, one that had been conserved from the [lower organisms] right up to man. When we found that mutation, we knew we had our gene. We were thrilled."

Vogelstein had a long relationship with the journal *Cell*. He had published numerous discoveries there and was on its board of scientific advisors. Vogelstein called the editor and suggested that his group's results be published in the same issue as Kolodner's research. Because the editors already had Kolodner's work in hand they decided to publish Kolodner's discovery first, on December 3,[7] and to publish Vogelstein's discovery in the next issue, two weeks later on December 17.[8] But, as a compromise, the journal editors agreed that the two papers could be released to the public and the press at the same time. It was a decision that many people in the scientific community took as a statement from the journal that it believed that the labs had ended up in a virtual tie.

It turned out that the mutation Kolodner and Fishel identified in their paper was a DNA anomaly of a type often seen in DNA, not the kind of mutation that can be passed from parent to child, but a common and benign genetic "typo." "No doubt we were in a hurry once Bert called us," says Kolodner. "If I had it to do over, I'd have taken some more time to publish, especially since we did find the mutations by the time the paper came out. But we were under extreme pressure, tremendous pressure. I'd never experienced anything like it in my entire career."

On November 29, Kolodner and Fishel received a call from Francis Collins suggesting they both come to NIH in Bethesda for a joint news conference with Bert Vogelstein. "How could we refuse?" Kolodner says, conceding that a press conference before the national media in Washington was certainly going to draw more attention than a local media event in Boston.

Indeed, the press offices of NIH and the Department of Health and Human Services trumpeted the find.

"This is a great day for science," hailed HHS Secretary Donna Shalala in a prepared press release. "The discovery of a 'cancer gene' is a textbook example of the kinds of payoffs we can expect when we invest in basic research."

Based on their surveys of families stricken with colon cancer, Vogelstein and Lynch told the media that they believed the HN-PCC gene accounted for as many as one in seven colon cancers in the U.S. "The faulty gene, believed to be present in more than one million Americans, may be the single biggest cause of inherited cancer," wrote Rick Weiss, on the front page of *The Washington Post*.[9] Like other medical journalists who attended the packed press conference, Weiss carried back a simple message: The new discovery, he wrote, "is the most significant advance yet toward understanding the molecular underpinnings of hereditary cancers." And like most others in the media, *The Washington Post*, while noting that *Cell* was publishing Kolodner two weeks ahead of Vogelstein, pretty much depicted the race a draw.

The intensity of the competition—both Kolodner's desire to go it alone, and Vogelstein's desire that his lab get the credit they had earned for the years of hard work—are surprising only to those unfamiliar with the strong personalities that dominate high-stakes science. Researchers and academics are as competitive as those engaged in other creative or entrepreneurial activities.

Both parties would admit that neither group could have found the gene without the other. "There's no way we could have found the gene so quickly without Richard's previous research," Modrich says. "And, of course, Richard knows he wouldn't have gone after the gene if not for the work that Bert had done. These guys are both good guys. I say, 'Congratulations' to them both."

The day after the NIH press conference Kolodner was a celebrity, appearing with Fishel on *Good Morning, America*, and quoted on front pages everywhere. Kolodner returned to Boston, a relative unknown no more. But Kolodner was plotting his next move. He and his collaborators had found three yeast versions of the bacterial repair genes. And only one of them was found in chromosome 2 of humans. The hMSH1 gene he and Vogelstein had just found was the human version of the bacterial mutS gene, but Fishel and another researcher, Michael Liskay, had also found another repair gene called mutL. They all figured there had to be a human version of that, too. "There had to be two more human genes somewhere," says Kolodner. "I was certain, especially since the ten families we looked at didn't have the chromosome 2 gene. There had to be more human repair genes. It was a hunch, but I figured it was a damn good one."

Kolodner took a few days off and then returned to the lab, as obsessed as he was before. And in Baltimore, Vogelstein told Ramon Parsons: "Let's get the other genes."

It was just about then that the last and most surprising of coincidences occurred.

Soon after Kolodner returned to his lab he received a call from researcher Mike Liskay, who had worked with Tom Petes in North Carolina. Now in Oregon, Liskay told Kolodner that he had cloned a mouse version of the other bacterial gene, mutL. He, Kolodner, and Fishel decided to use it as a probe to fish out the human version, if it existed. By the middle of January, the two labs had found another human gene, located on chromosome 3, and they dubbed it hMLH1. Immediately, Kolodner began analyzing families, finally getting several useful ones from a medical research team he'd heard about in Sweden. By late January, he began hunting for mutations in hMLH1 in the Swedish family. "This time I was going to be certain I had a strong set of mutations before we went to print," Kolodner says.

In Baltimore, Vogelstein knew he probably only had a few weeks before Kolodner found the human version of the bacterial repair gene, mutL, if it existed. For a month, the lab tried doing what Kolodner had done with Liskay's help, but they had no luck plucking out the human version of mutL. So they decided to use a different strategy. "Bert has great instincts, it's part of what makes him a great scientist," says Kinzler. "He came in the lab one day and said he had this idea where we could find the gene, and he was right." Vogelstein's idea was to search well-known computerized databanks at NIH and elsewhere that contain a list of all published genes or partial genes. His goal was to see if anyone had ever accidentally cloned all or part of mutL homologs without understanding their importance.

"We couldn't believe what we found," Vogelstein says. In fact, a group of scientists in nearby Gaithersberg, Maryland, at a lab called The Institute for Genetic Research, or TIGR, had found bits of the gene a year earlier. Vogelstein had met the director of TIGR, a controversial scientist named Craig Venter. The previous year Venter had generated headlines himself by developing a technique to quickly isolate fragments of human genes. In fact, while working at the NIH Venter was cranking out hundreds of these fragments, which he proposed to patent. When that idea stirred up a storm—by scientists who said such small bits of DNA couldn't be patented—Venter went off and formed a private research institute funded by a new biotechnology company in order to exploit his gene-isolating technology.

The company, called Human Genome Sciences, Inc., or HGS, was instantly backed by a large stock offering and $125 million from SmithKline Beecham, a drug maker that believed the gene fragments would eventually help the company identify new disease pathways.

Ken (Kinzler) called Craig up and asked him if he had human mutL homologs. " 'Yeah,' Venter says, 'we found three of them but they really belong to HGS. I better call Bill,' " Kinzler recalls. Vogelstein and Kinzler knew Bill was William Haseltine, chief executive officer of HGS, and a giant scientist in his own right. Back in the late 1970s, Haseltine made a name for himself at Harvard studying DNA repair. In the 1980s he jumped into AIDS research, helping to identify many of the important genes that make up the AIDS-causing virus, HIV. In 1993, he was recruited by Venter and some venture capitalists to run HGS.

Venter called Haseltine and told him: "Bert's in a helluva race and he needs our help." A broad-shouldered, extremely confident man who enjoyed socializing with a celebrity crowd in Boston, New York, Los Angeles, and Washington, Haseltine jumped at the notion of helping Vogelstein out, and for the attention finding the gene might give to his fledgling company.

Moreover, Haseltine knew exactly what Vogelstein wanted. Only several days before Haseltine had called Kolodner, whom he knew from his days in Harvard. Apparently, a researcher at HGS had stumbled across the mutL homolog during one of the company's numerous gene-fragment searches. Haseltine knew the importance of mutL as a mismatch repair gene and he told the researcher to find as many similar genes as possible. Ultimately, HGS found bits of three different genes that each looked as if they belonged to the mismatch repair family.

It was a few days after the New Year in 1994 that Haseltine called Kolodner, "I'd seen the colon cancer papers, and I figured we had genes he was looking for," Haseltine says. "I told Richard I had these homologs, and that he ought to come down here and take a look. Instead, Richard just told me he was going skiing. Now I said, 'Richard, really, I think you're going to want to come down here.' And he tells me again that he really needs to go skiing."

Kolodner laughs, remembering the story. "We had the gene by then," he says. "I wasn't interested in collaborating with Bill, and I didn't want to make him think he had something important."

But Kolodner didn't expect Haseltine to be talking to Vogelstein, at least not so soon. But within days of realizing what Haseltine had, Vogelstein and Kinzler were in a car driving the hour to Gaithersberg, Maryland. They arrived with a bucket of ice, inside which were cell lines containing DNA from the colon cancer tumors. Soon afterward, the HGS and Baltimore labs had found a human gene on chromosome 3 that likely was the cause of colon cancer in the many families that were not linked to chromosome 2.

Back at the lab, the researchers were ecstatic. "We had been struggling to find the genes for more than a month and here Bert goes shopping for them down the street," says Ramon Parsons. Adds Nick Papadopolous: "We called it 'Clone by phone.'"

The entire community of gene-hunting scientists were amazed by Vogelstein's strategy and by the serendipitous existence of the genes. "To us, it showed how far we had come in just a few years in elucidating the human genome," says Curt Harris, the National Cancer Institute scientist. "Here Bert walks over to Haseltine's building and walks out a few hours later with his genes. It was the fastest way to get the genes, and Bert figured out how to do it. You've got to give him a lot of credit. It was a very clever, very lucky turn of events."

As it had three months earlier, Vogelstein's swift acquisition of the gene was soon to raise eyebrows among his friends, and carping from some other scientists.

By the middle of February both Liskay and Kolodner and Vogelstein and Kinzler had found mutations to hMLH1 in the colon cancer families. And once more in back-to-back research reports, the two teams reported their findings, once again to the national press. *Science* magazine itself reported: "The study of colon cancer genetics is red hot. For the second time in three months, two teams of researchers have raced to uncover a gene involved in a hereditary form of colon cancer, and for the second time the race ended in a dead heat."

12: The Mother of All Tumor Suppressors

Staring at his computer terminal on a cold evening in late January 1994, Sasha Kamb came as close to shouting *Eureka* as is possible for a sober, modern-day scientist. "This is it," Kamb said. "Jesus, this is incredible." What Kamb did later that night and the next few days would set off one of the more bizzare bits of theater to emerge from the hunt for cancer genes.

On the computer screen before Kamb was a long list of strangely named genes culled from the world's largest archive of DNA data. Housed in a mammoth computer outside Washington, D.C., called Genbank, this "library of life" is an extraordinarily powerful gene-sleuthing resource. Moments earlier Kamb had electronically entered the library by punching into his keyboard the sequence of letters that spelled out the partial structure of a gene he and his colleagues at Myriad Genetics had been stalking for eighteen months. Evidence gathered by colleagues at the University of Utah for the past six years led him to suspect the gene he was chasing helped trigger melanoma, a brutish form of skin cancer. With one more keystroke, Kamb dispatched the gene's partially known code of dozens of assorted A's, T's, C's, and G's from his lab in Salt Lake City through the Internet and into the library's database of 85,000 genes, every nonplant gene ever discovered. Like a police detective seeking to unmask a criminal by comparing a suspect's fingerprints to a collection of known prints, Kamb used Genbank's automated scanning program to uncover similarities between the known letters of his elusive gene and any other gene whose identity was entered into the databank's collection.

This procedure is called a "blast" search, and it is similar to the computer-hunting technique used by Bert Vogelstein a month earlier to track down the second HNPCC gene. A blast search is an especially clever trick because it exploits a powerful law of genetics: As was seen with the mismatch-repair genes, many human genes are the descendants of genes that first emerged in prehuman organisms. If some other scientist, for instance, previously had isolated a gene instrumental in the blood vessel formation of a worm, that gene, its sequential code of letters, its biological function, and the discovering laboratory would be listed in Genbank. Genes that carry out critical jobs with a special efficiency in lower species are "conserved" through the hurly-burly of hundreds of millions of years of evolution. These genes are like precious heirlooms inherited by the human genome from unnumbered generations of biological ancestors. Sometimes the human genome acquires the animal gene whole, but more often a new segment is added or another is changed or subtracted.

A blast search through Genbank, therefore, might provide a scientific shortcut for uncovering a gene's identity and purpose. If, by chance, some significant stretch of Kamb's gene were similar to a worm's vessel-forming gene, it might mean Kamb's human gene had a similar design and function in humans. A mystery can be solved in seconds, the amount of time Genbank takes to survey its catalogue and return with a roll call of potentially related genes, those genes with the greatest similarity, or homology, at the top of the list.

That's why every few weeks or so, as Kamb and his team unraveled more of their gene's DNA sequence, they routinely shipped the expanded code to Genbank. Usually, Genbank's automatic scanning program replied with a long column of obscuresounding genes from species as diverse as yeast, bacteria, fruit flies, nema-

todes, mice, macaques, or orangutans, in which some small stretch
of the DNA from the library's gene overlapped by coincidence
with the partial sequence Kamb had exposed in his gene. The
genes' homology always was too small to be considered a legitimate match.

But not this time. At the very top of the Genbank roster inno-
cently sat a human gene with an especially esoteric name. Kamb
blinked a few times as he looked at the screen, not comprehending
right off what he was seeing. The partial sequence of his gene and
the Genbank gene weren't just similar, they were identical, forming a match of the sort Kamb had never seen before from a blast
inquiry. There was no way around it; someone already had found
Kamb's gene. In fact, according to the brief annotation accompanying the gene's listing, its sequence had been deposited in the library only several weeks earlier by gene-hunting researchers at the
prestigious Cold Spring Harbor Laboratory on Long Island's north
shore. But from the limited information available from the Genbank catalogue, Kamb figured that the Cold Spring Harbor lab
didn't yet know what he knew, that their gene was quite possibly
the most important cancer-causing gene yet found.

Sitting alone in his office peering at the results of his "blast hit,"
Kamb felt the first inklings of the drama to come in the following
weeks. He was excited that suddenly in the flash of an electronic
data search he had solved many of the mysteries surrounding his
previously unnamed and unidentified gene. But he was disappointed because another lab provided the critical clues. Also, he
was especially apprehensive because he realized he had joined a furious race that moments before he didn't know existed.

Up until then Kamb had been mildly skeptical about the importance of the melanoma-causing gene he was pursuing. After all,
he had no idea what role this gene played in a cell, or whether the
problem it caused was unique to skin cancer, or even to most cases
of the cancer. For months, Kamb had been puzzled by experiments
in tumor cells from other types of cancer showing that large
swatches of DNA often had become mutated in the same region
that he believed was home of the unidentified melanoma gene.

According to the Genbank annotation, the Cold Spring Harbor lab, run by British scientist David Beach, said the gene monitored a critical junction in the life cycle of a cell. In itself, that discovery was quite important, even though Kamb had been previously unaware of the gene's discovery or its publication by the Beach lab two months earlier in the esteemed journal *Nature*.[1] Beach's team said the gene's protein, which it dubbed p16-ink4, was a "cyclin-dependent kinase 4 inhibitor." Kamb understood its importance right away.

"This gene isn't just involved in melanoma; it's a major tumor suppressor gene," Kamb whispered to himself. "And Beach doesn't know it. Or at least, he hasn't said so."

Kamb took a deep breath, rechecked the data supplied by Genbank, pushed back his chair and made a phone call to his boss, who had gone home for dinner several hours earlier. "What's up?" Mark Skolnick asked, knowing instantly from Kamb's tone, so different from his normally relaxed manner, that something important had happened in the lab.

"Somebody's already found our gene," he said. Before Skolnick could respond, he added, "You won't believe this. Genbank says it's a cell cycle regulator listed just a few weeks ago by Beach at Cold Spring Harbor. It explains a lot. I think it means we've gotten hold of more than a gene that causes melanoma. We have a tumor suppressor here, maybe a very big one."

Skolnick told Kamb he'd be right over. Kamb hadn't eaten dinner; he hadn't expected the blast to take much time, and early evening had now passed into night. He asked Mark to bring him a sandwich. Angela, Mark's wife, stuffed some ham, leftover from their dinner, between two slices of bread, Mark wrapped it up and sped off down the hill to Myriad's laboratories on the eastern fringe of University of Utah campus.

Between bites Kamb told Mark what had happened. First off, he explained that the Cold Spring Harbor lab had found a gene that expressed a protein, called p16, that acted as if it were a natural "off-switch" deep within a cell's nuclear machinery. Its task was to inactivate an enzyme called cyclin-dependent kinase 4, or CDK4, that spurs a cell to reproduce itself. The p16 protein carried out this task by somehow chemically embracing CDK4 in a bear-hug that stifles cell division.

The two genes, the one that makes the CDK4 enzyme and the one that makes p16, act like counterbalancing gears in an automobile transmission. When the CDK4 enzyme is active, the cell's duplicating engine runs at full tilt. When the p16 protein is expressed by its gene, the engine goes into neutral gear and cell division is arrested. If the gene that makes p16 is damaged or defective, the engine revs out of control and cells divide unchecked until their progeny accumulate into a massing of cells that become a tumor.

As Kamb explained this, Skolnick began to understand the implication of such a gene immediately. He hugged Kamb. They exchanged smiles, and immediately began plotting their next move. Skolnick was elated. After six years of frustration he'd found the gene! But he also was worried. He and Kamb agreed it was bad luck that the gene's identification already was published, even though, they had to admit, its discovery and listing in Genbank made their find possible. But its publication might cause Myriad a problem getting a broad commercial patent for the gene. And in order to create a proprietary diagnostic test and therapies based on the gene, a strong patent claim was critical. Skolnick and Kamb knew Beach only by reputation as someone who was very smart, very ambitious, and backed by a brilliant band of researchers he could quickly marshal if he realized he was in a race with another lab. While they suspected Beach already had filed a patent for the gene, they hoped their application would be stronger.

Their job was clear. Beach had found the gene and, importantly, he had described its normal function. But he had not shown that, in mutated form, the gene could trigger cancer. At least, there was no published indication that skin cells containing the cell cycle gene in defective form were likely to become cancerous. Kamb, however, had had hints. Since the previous summer, Kamb and his group had sifted through the region of DNA containing the gene in cell lines of other types of cancer. These cell lines were cancer cells isolated and removed from cancer tumors and then "immortalized," prodded through various techniques to reproduce endlessly in laboratory petri dishes, then stored in freezers for removal at a later date when they could be thawed and studied. Such immortal cell lines were highly prized commodities, and researchers who grew them were cultivated and wooed by the molecular biologists. Skolnick had several well-guarded cell line suppliers: doctors and researchers in medical centers around the country. In recent months these cell line growers mailed him samples packed in dry ice. Kamb's lab then extracted the DNA from the tumor cells and tested whether the region that contained the suspected melanoma-causing gene was also altered in some fashion in other cancers.

By the time he had performed the blast search earlier that evening, Kamb already had preliminary evidence strongly hinting that the partial bit of gene he had grabbed hold of was often mutated in tumor cells removed from patients with cancers of the brain and the kidney. He and Skolnick decided the lab should momentarily put aside its furious efforts to find a breast cancer gene and concentrate the lab's skilled personnel into surveying tumor cell lines from as wide a range of cancers as was possible to do quickly. If, as they hoped, Beach's gene was altered in many types of tumors, then they had the makings of a major discovery, a tumor suppressor gene to rival in importance any found to date.

"Mark was pleased, very excited," Kamb recalled in a conversation a few months later. "He'd been looking for the melanoma gene for years, and while he was convinced of its importance, it really wasn't clear how big a deal it would be. What I'd told him suddenly upped the stakes. We definitely had something important. He understood that. Yes, he was very pleased."

He also was a nervous wreck. The tumor suppressor discovery, if true, *was* an unexpected gift, and Skolnick believed he had to move swiftly to exploit it. Myriad Genetics, although only two years old, quickly was gobbling up cash from the private investors who had been enticed by the company's promise of finding new powerful gene-based cancer products. Like many newborn biotechnology companies, Myriad's attraction was its unblemished potential. It had no marketable products yet. In the spring, Skolnick and his partners planned to hit the road Wall Street-style, seeking fresh funding from investors in the United States and abroad, money crucial for making long-range plans. A coincidental discovery certainly would make Myriad's story more thrilling, help spice up its sales pitch, and enhance the company's chances of survival. But for Skolnick there were bigger stakes involved.

Skolnick wasn't a molecular biologist—a heroic lab bench jockey like Vogelstein or Collins who could devise new ways to dissect bits of DNA. Skolnick never tried to fool himself into believing he was. His strength, for which he was rightly pleased, was deciphering inherited patterns hidden in the Utah genealogy. That was an extraordinary contribution, and biologists such as Vogelstein were constantly phoning Skolnick and his population geneticists, hoping to mine their rich lode of family data. Already, these data had helped researchers uncover the existence of genes involved in cancers of the colon, breast, and skin and a host of other common ills.

But while Skolnick and his University of Utah staff were adept at mapping a gene to its imprecise location on a specific chromosome, actually hooking the gene itself—the true object of all this, after all—had been largely beyond their grasp. With the $10 million Myriad Genetics had secured from private investors, and the expectation for millions more, Skolnick had the resources—the money, space, equipment and, most important, the technical talent of people such as Sasha Kamb—to be not just a player, but one of a handful of the world's preeminent gene hunters, able to track down and bag the biggest genetic game: the genes that cause humankind's most intractable and deadly diseases.

With Myriad's well-appointed labs, Skolnick first planned to bootstrap back into the center of the gene-hunting arena by discovering a gene that predisposes people to melanoma, a killer if caught too late, but curable if diagnosed early. A predisposing gene for skin cancer would be an immeasurable public health contribution. Almost 7,000 Americans die each year from skin cancer that has metastasized, murderously colonizing the liver, bone, or brain. Another 32,000 cases are found each year in a stage so advanced they must be treated with aggressive surgery, chemotherapy, and radiation. Even then, chances a patient will survive five years are slim.

Skolnick believed unearthing the melanoma gene would change the odds. If he could prove that most, or many, melanomas arise because people inherit a gene defect, then it would be possible to create a blood test that would identify who was at high risk for the disease. In numerous talks and published articles, Mark argued that gene carriers, forewarned of their increased danger, would be highly motivated to guard their skin against excessive exposure to the sun's rays. They might also have their skin closely examined for dysplastic nevi, M&M-sized moles that can progress under certain circumstances into killing tumors. Parents might have better reason to protect children born with the damaged gene, especially since doctors believe a severe sunburn early in life can unleash the slow, relentless molecular changes characteristic of skin cancer.

There was, of course, another thing, too. Myriad's success could make Skolnick a multimillionaire. When the company was formed in 1991, Skolnick purchased 2,368,000 shares of "founder's stock" at $.0002 per share, a typical "paper" bonus accorded the founding officers of high-technology company start-ups. A year later he purchased another 1,184,000 shares for the same price, and he also was granted an option to buy another 1,065,600 shares at $.0004 a share at a later date.[2] If Myriad ever went public and its stock traded as high as just $1 a share, Skolnick's holdings would be worth $4.62 million.

But before he could fantasize about building a vacation home on the ski slopes above Salt Lake City where his son, Josh, could ski-board out their front yard, or buy a pied-à-terre in his wife Angela's beloved northern Italy, Myriad had to find a gene to call its own. The one he believed predisposed people to skin cancer seemed the best bet. Prior to late 1992, however, few scientists and doctors were convinced such a gene existed despite the claims made by Skolnick and his colleagues at his small University of Utah facility. But by probing his famous Utah database, Skolnick's team had uncovered numerous large families where melanoma was unusually common, too frequent to be due solely to chance. In the fall of 1992, Skolnick's band of university researchers finally mapped a common gene in these families to the short end of chromosome 9.[3] At almost the same time, two other well-respected labs, one in Chicago and another in Cambridge, Massachusetts, also placed the gene in the exact same spot. All three teams, as well as about four or five others, entered into a scientific sprint to isolate the culprit gene's exact identity and function, and prove that people who inherited it in altered form are best advised to avoid intense sunshine.

At the same time that Kamb undertook his search to isolate the melanoma gene's exact molecular structure, across the continent in laboratories on Long Island's wooded northern shore, David Beach was trudging toward the same bit of DNA from a completely different direction. Beach already had staked himself a giant-sized reputation as an upcoming molecular biologist. His field of inquiry wasn't human cancer, at least not directly. Instead, he and his lab were expert at charting the complex terrain of the cell cycle, a series of reactions in the cell's nucleus where proteins and genes perform a biochemical do-si-do, sharing and exchanging partners in a precise yet complicated dance, their coupling and uncoupling signaling a cell when to divide, to stop dividing, and divide again, over and over. Beach studied the cycle in the cells of yeast, microorganisms whose biology is simpler to dissect than that of humans, but whose proteins and genes had been conserved through the eons of evolution and now often resided in a surprisingly similar design in human chromosomes.

Beach arrived at Cold Spring Harbor lab as a postdoctorate researcher in 1982, and quickly grabbed the attention of the lab's leaders for being innovative and seemingly fearless. By 1994, Beach had made numerous impressive insights, had his own lab, and was widely accepted as one of the rising stars at an institution with no shortage of brilliant scientists.

The late Vice President Hubert H. Humphrey and his wife, Muriel, shown here on the 1968 presidential campaign trail. Humphrey died almost ten years later from a bladder cancer that researchers now know was already stirring inside him during his quest for the presidency.

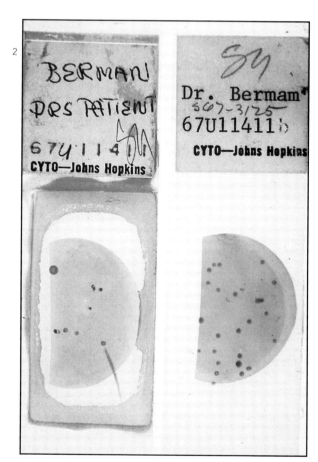

Two microscope slides of bladder cells from a 1967 urine sample taken from Humphrey. The slides were stowed away in the drawer of a suspicious pathologist for 26 years, until gene researchers in 1993 identified the genetic cause of the vice president's bladder cancer.

3

Dr. Barbara Weber, a geneticist at the University of Pennsylvania School of Medicine. In a fateful meeting in August 1992, at the University of Michigan, Dr. Weber used a breast cancer gene test to save a young woman from unnecessarily undergoing surgical removal of her breasts to protect herself from the breast cancer that stalks her family.

4

Mary-Claire King, the geneticist who led a twenty-year crusade to uncover the existence of the breast cancer gene, BRCA1. King is pictured here in her lab at the University of California, Berkeley, sitting in front of a poster describing the effort to identify children taken from parents of dissidents during the fascist regime in Argentina in the 1970s. King used genetics to help re-unite some of the children and their families.

5

Francis Collins, director of the National Human Genome Research Institute. In 1990, while still at the University of Michigan, Collins joined with King in a famous four-year collaboration to isolate BRCA1.

Mark Skolnick of the University of Utah and Myriad Genetics, Inc. In 1994, Skolnick led a team of scientists who isolated BRCA1, ending one of the fiercest and most publicized of gene-hunting races.

Kenneth Kinzler *(left)* and Bert Vogelstein *(right)* codirectors of the Johns Hopkins Oncology Center in Baltimore. The researchers stand beside a makeshift robot built from washing machine parts by Kinzler that was used to identify mutations of the gene p53 in colon cancer cells. The 1988 mutation discovery was the first evidence in human cells that a defect in p53 can transform a normal cell into a cancer.

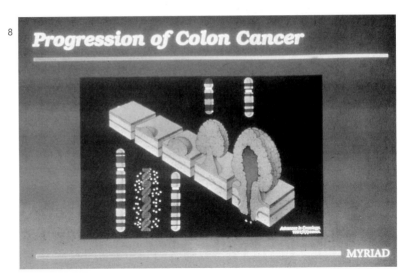

An illustration depicting Bert Vogelstein's insight showing how a normal colon cell turns into a cancer after accumulating several gene mutations.

In 1993 and again in 1994, cancer-causing genes identified by Vogelstein and others were named by the prestigious journal *Science* as "Molecule of the Year."

11

Jane Green, a geneticist at St. John's University Health Science Centre in Newfoundland. Green's exhaustive research into a Newfoundland pedigree helped researchers identify the existence of a colon-cancer gene.

Albert de la Chappelle, seated between his colleagues Lauri Aaltonen (left) and Paivi Peltomaki, of the University of Helsinki in Finland. De la Chappelle's research team identified families in Finland whose colon cancer was caused by inheriting a "mismatch repair" gene. He joined with Vogelstein and others to identify the gene in 1994.

12

13

Richard Kolodner of the Ludwig Cancer Research Center, La Jolla, California. In 1994, while at Harvard University and the Dana Farber Cancer Center in Boston, the little-known Kolodner raced Vogelstein and others to isolate the first human mismatch repair genes.

Several especially lively members of Vogelstein's research team, seen here in their alter-egos as the rock band *Wild Type*. Trained as a classical pianist, Vogelstein *(third from left)* is the rock group's spiritual leader and keyboard player. The members of the band, which played frequently at a local Baltimore bar, are *(left to right):* Robert Casero Jr., Ph.D. (who directs his own laboratory); Chris Torrance, Ph.D.; Vogelstein; Kenneth Kinzler, Ph.D.; Elie Carson, doctoral candidate and lead singer; and Pat Morin, Ph.D., song-writer and singer.

14

Mark Skolnick *(left)* and Sasha Kamb *(right)* of Myriad Genetics, Inc. Kamb found that mutated forms of the gene p16 were involved in causing many different types of cancer. Myriad's discovery of the gene's cancer-causing role became the center of a controversial press conference in 1994, highlighting the intense competition among scientists to claim patent rights for the genes.

15

16

Donna Shattuck-Eidens (seated) and Skolnick. Shattuck-Eidens led the team that finally found BRCA1 in the summer of 1994.

Mary-Claire King *(right)* and colleagues at the University of California, Berkeley, in search of the breast cancer gene.

17

Allen Oliff (*left*) and Edward Scolnick of Merck. Scolnick joined the New Jersey pharmaceutical giant in 1978 and over the years inspired the first concerted effort, led by Oliff, to develop a drug against a cancer-causing gene.

18

Patrick Walsh, director of the Brady Urological Institute at Johns Hopkins Medical Institutions. Walsh, a surgeon specializing in treating prostate cancer, collected the families that helped identify the presence of an inherited prostate cancer gene in 1996.

19

Barbara Biesecker (*right*), a genetics counselor, conducting a mock interview with a breast cancer gene family at the National Human Genome Research Institute. Biesecker was part of the team that helped counsel the Michigan "family 15" in 1992 and now helps design counseling programs for families that carry cancer-causing genes.

20

Kamb and Skolnick barely registered on his scientific radar screen. Yet, in the spring of 1993, while Kamb's Myriad lab was gearing up for the last lap of the melanoma gene chase, Beach's lab was making a series of impressive discoveries of previously unknown cell-cycle proteins, one of which he called p16. Apparently thinking his team alone was onto the gene's existence, he didn't rush to publish his finding, waiting until December 1993, to announce it in *Nature*.[4]

Months later, sitting behind his desk in a second-floor corner office with a bright and sweeping view of a marshy inlet opening onto Long Island Sound, he conceded that he'd made an error. "We should have published [discovering p16] sooner," he said. Staking an earlier claim to the gene might have blocked the competition to come. "I just had no reason to suspect what was going to happen. I'm still astonished."

Only after the *Nature* article was released was the p16 gene, including its full DNA sequence and its function, finally listed in Genbank's register of genes, where its presence was picked up several days later by Sasha Kamb's blast. Although Beach's p16 find was big news to the coterie of scientists piecing together the cell cycle's complex machinery of biochemical valves and gears, Kamb and Skolnick paid it no heed until Kamb saw its description burst across his computer screen. Realizing they suddenly were pitched into a race for time, Skolnick told the Myriad lab staff to get ready for weeks of marathon workdays. Over the next few weeks, he and Kamb probed many more tumor cell line samples dispatched to Utah from a research team at Memorial Sloan-Kettering Cancer Center in New York, scouring the site where the p16 gene was located. By early March, Myriad's staff had analyzed 290 tumor cell lines, and they found in 46 percent of the cell lines wide areas of DNA missing or damaged where the p16 gene should have been intact. The evidence was as profound as it was unexpected. p16's damage in these cell lines could only be due to one reason, Kamb and Skolnick decided: in defective form the gene could no longer stop the cycle that caused the cell to divide. Thus, it unleashed the type of unfettered cell cloning typical of a massing tumor in a potentially malignant cancer. Around Myriad, Skolnick called the gene "the mother of all tumor suppressors."

Skolnick personally packed up the data and carried it under his arms to the U.S. Patent Office's outpost in San Francisco, claiming that Myriad Genetics had uncovered a new cancer-causing gene. This he knew was going to be tricky. There was no doubting that Beach had found the gene first and that his employer, Cold Spring Harbor Laboratory, likely had filed a patent for it. Moreover, Beach also recently had founded a company, based in Cambridge, Massachusetts, called Mitotix, Inc., to develop cancer diagnostics and therapeutics based on his cell cycle protein discoveries. Skolnick was convinced Beach would lay claim for the gene.[5] But he hoped Beach hadn't yet found any evidence linking p16 directly to cancer. To make certain no one confused the two research claims, Kamb christened the gene anew, calling it MTS1, for multiple tumor suppressor 1.[6]

Skolnick was desperate to know what Beach knew about p16 and cancer, but he didn't want to tip his hand. Once alerted that his p16 gene potentially was one of the biggest cancer gene finds yet, Beach would certainly be able to produce the same cell line data found by Myriad. With his influence and reputation, Beach might even be able to beat Myriad to press, perhaps wounding Myriad's patent position. The young field of cancer genetics was too small and competitive for Skolnick to trust asking anyone close to Beach for fear that might alert the Long Island team. Several months earlier, Myriad had hired a small public relations firm from Boston, Feinstein & Partners, to begin a publicity campaign when the melanoma gene was found. Peter Feinstein, the company's founder, specialized in representing emerging biotech firms. He knew the reporters who covered the industry, as well as those in the investment community who should be apprised of biotech happenings, and he was good at communicating the importance of breaking research news. But in still another coincidence, Feinstein also represented Mitotix, Beach's new cell-cycle company. In the small world of biotechnology public relations, such an overlap was not that surprising. But, as a result, Skolnick kept Feinstein and his employees in the dark, refusing to say exactly what the discovery was that they were being asked to promote.[7]

By the second week of March, Kamb's research team was convinced it had found enough evidence to publish a scientific paper declaring that Beach's p16, renamed MTS1, was a tumor-causing gene. To their dismay and surprise, however, they were unable to show conclusively that the same gene, if inherited in mutated form, predisposed people to melanoma. When Kamb looked at the DNA of people in families with melanoma he couldn't find any consistent mutation in MST1 that could account for their shared risk.

"It was the most puzzling of things," Kamb said, beginning to wonder anew if there was such a thing as a gene that propelled skin cells toward cancer. "Mark, of course, was convinced MTS1 predisposed people to melanoma. The mutation seemed to be hiding someplace obscure, and it wasn't going to be very easy to find it, and we certainly weren't going to find it fast."

Kamb and Skolnick briefly considered waiting to publish news about MTS1 being a tumor suppressor until they found evidence related to melanoma. But after a short debate, they decided the tumor suppressor find was too hot to keep.

On March 17, Kamb sent the research paper to the elite journal *Science*. Personally talking with *Science*'s editor, Daniel Koshland, Kamb and Skolnick shared some of their worries about how easy it would be for the competition to eclipse them. He urged Koshland to take great care in choosing scientists to conduct a peer review, specifically requesting that neither Beach nor anyone at Cold Spring Harbor get an early peek at the results. Skolnick decided to drive a tough bargain. He told Koshland that unless he agreed to publish the paper quickly, faster than the six weeks it normally took to process an important advance, Myriad would seek to publish it elsewhere. Koshland was excited about carrying the news of a major cancer gene discovery in his publication and agreed to move quickly. He didn't want rigid editing strictures to cause Skolnick to publish the paper in the pages of one of his main competitors, *Nature* or *Cell*.

Still, given the small and incestuous nature of the gene-hunting community, it was probably inevitable that the news would dribble out. What was surprising was that it occurred so quickly and that the source of the leak was Myriad. Back in the second week of February, shortly after the blast hit, one of Myriad's top gene-hunters, Donna Shattuck-Eidens, was setting off to attend a major gathering of cancer gene investigators at a resort in the Sangre de Cristo mountains above Santa Fe, New Mexico. It was a private weeklong meeting that had been planned a year earlier. In checking the list of speakers, Shattuck-Eidens saw that Beach was scheduled to talk about p16. "See what he's saying," Skolnick suggested. "Just be careful."

Right off, Beach's behavior made Shattuck-Eidens suspicious. For one, he failed to appear at his scheduled time on Tuesday morning, February 15. The meeting's organizers rescheduled Beach for that evening, but when he finally arrived he postponed his talk until the next day, saying he was exhausted from the trip. The next morning he began his talk by explaining that because of heavy traffic getting to the airport he had missed his flight the day before by just moments. Beach then joked that being minutes late in molecular biology these days could be equally disastrous; coming in second with a published finding meant you didn't get credit for the work even if your lab actually had made the discovery first. Beach didn't have to say what he meant. In the insular world in which these scientists moved, it was well-known that Beach had lost another race the previous autumn. Several labs, including the one run by Bert Vogelstein, reported an important development related to another cell-cycle gene called p21 found by Beach the previous spring, about the same time he found p16. The labs all published that p21 was critically involved in causing colon cancer by interacting with p53, the most important tumor suppressor yet uncovered. Beach had similar data but Vogelstein and the others had gotten it out faster, reaping the acclaim that Beach and his colleagues believed was theirs.

"This is pretty pathetic; he's going to get scooped again," Shattuck-Eidens thought. But then, she worried, maybe she was wrong. The title of Beach's talk specifically said it was to focus on p16, but he barely mentioned it. Running into Beach later, Shattuck-Eidens introduced herself. She told Beach she was looking for cancer genes, in particular the one underlying breast cancer, and, in a query that would haunt her later, she asked Beach whether he thought p16 was a tumor suppressor. She asked Beach if he knew where it was located in the genome, and he said his lab had mapped it to a particular spot on chromosome 11. Shattuck-Eidens knew that was wrong, that Kamb had found the gene only because he was exploring chromosome 9. "Is he lying or is he just wrong?" she wondered. Beach asked her whether Myriad had any evidence of a cancer gene in the area, and she said she didn't know but would get back to him if they ever did. When she asked him why he hadn't discussed p16 specifically in his talk, Beach charmingly changed the subject, suggesting that they might work together to find a cancer gene if she ever found evidence of one in the p16 region. A few days later, Shattuck-Eidens returned to Myriad, passing along what little she gleaned. "He seems more interested in p21," she told Skolnick. Skolnick didn't believe Beach could have mismapped p16. He, too, wondered if Beach was purposely sending out misinformation. The report certainly didn't calm him any.

At the New Mexico meeting Beach had honestly believed that p16 sat on chromosome 11. But he had good reasons for being cryptic. Back in January, a few weeks after the *Nature* article was published, Beach received a call from Curt Harris, who ran the National Cancer Institute's prestigious molecular biology labs on the National Institutes of Health campus in Bethesda, Maryland. Harris didn't know Beach personally. But he saw an opportunity in the *Nature* paper. Harris was the NCI's principal authority on tumor suppressor genes. His lab was one of the leaders in mapping the pathway by which the p53 gene transformed a quiescent cell into a raging tumor. When he read Beach's *Nature* article, he was convinced that p16 might be a p53-like gene, potentially a new tumor suppressor. "It just made sense," he later said. He called Beach and told him his suspicion. Moreover, he made an offer Beach could hardly resist. Harris had plenty of tumor cell lines, and, as important, he knew how to scan them for gene damage. Beach and Harris agreed to collaborate. Harris told Beach he believed others might soon get the same idea. "If you knew tumor suppressors, the idea that p16 might be one really wasn't that brilliant," Harris later conceded. "We had the know-how and the materials, so we decided to put aside what we were doing and do the experiments on p16. I felt we should get at it as quickly as possible, and Beach agreed."

By the time Skolnick had sent off Myriad's data to *Science* in mid-March, Harris's lab was seeing essentially the same results as Kamb. The p16 gene was damaged in almost half the cell lines tested. Beach and Harris decided to submit the experiments for publication. Then, a few weeks later, Beach received a disturbing call from Bruce Stillman, who had taken over as director of the Cold Spring Harbor Laboratory a few months earlier when Jim Watson had stepped down. Stillman had just gotten off the phone with Skolnick, who had made a very curious request.

Cold Spring Harbor Laboratory was planning a major, week-long symposium in early June that was expected to attract dozens of researchers from around the world representing every molecular biology laboratory of note. The Cold Spring Harbor get-together was widely considered a prime event, the kind of all-star gathering in which reputations could be made or broken, rumors could be sown, confirmed, or quashed, and fortuitous collaborations forged. Speakers included Mary-Claire King and Francis Collins, among others. Bert Vogelstein was giving a keynote address. Every morning for seven days, there would be talks in which scientists presented their latest experiments to a packed auditorium of peers. More research was presented again in the late afternoon, and again after dinner more presentations lasted late into the night.

The scientists were housed in dormitories on the campus and en-couraged to freely share their work, ideas and criticisms during the formal talks, meals, at special wine-and-cheese receptions, at an open bar outfitted with a pool table and darts, or in leisurely walks along the sandy shoreline. The agenda had been set months earlier when researchers were required to send in abstracts summarizing their talks. Cold Spring Harbor's staff then reviewed the research, choosing several dozen of the hundreds of entries deemed most important for oral presentations. The rest were relegated to less-impressive afternoon "poster sessions," in which written descrip-tions of several dozen experiments were posted on walls in adjoining rooms, the authors standing alongside. Researchers strolled thought the rooms, reading the descriptions and asking in-vestigators questions.

Skolnick originally had sent in a paper describing the joint ef-forts of Myriad and his university group to track down the melanoma gene. But now with MTS1 Skolnick believed he had something explosive that was certain to grab the attention of his peers, perhaps one of the most important new pieces of research of the year.

Skolnick called Stillman in early April, soon after *Science* told him it would be publishing the research shortly, perhaps within two weeks or so. Feeling it now was safe to share the MTS1 story, Skolnick asked Stillman to make room for his lab's presentation in the group of talks that would discuss developments in the cell cy-cle. In essence, he shared with Stillman details of the research that was still weeks away from being released publicly, asking Stillman to keep the information private. It was a gamble, but Skolnick be-lieved it worthwhile. It would ensure his group credit for a major advance among his peers, recognition he knew would help Myriad in its efforts to build all-important collaborations among the big hitters in the field. But shortly after Skolnick spoke with him, Stillman passed the curious news along to Beach.[8]

Beach wasn't certain what it meant. But then he remembered his brief conversation six weeks earlier with Skolnick's colleague, Shattuck-Eidens. She specifically had asked him about p16, he re-

membered. He called her up and, after some perfunctory comments, pointedly asked, "Is Myriad's melanoma gene the same
gene I'm interested in?"

"Oh, David, you're interested in so many genes. How would I
know?" she asked. Beach persisted. Is Myriad publishing a paper? Is
it already in press (submitted and accepted for publication)? At
first, Shattuck-Eidens fended off his questions, telling Beach there
was no paper planned. But, by conversation's end, she acknowledged that a paper had been submitted to *Science*, although she
didn't tell him its contents. "Is it p16, is it my gene?" Beach pushed
once more. "I can't tell you," Shattuck-Eidens finally said. "Really,
you've got to talk to Sasha or Mark, I can't say anything more."
Beach asked her to pass him along to Sasha.

She put down the phone, and hurried the 30 feet down the hall,
past the labs and into Kamb's cramped office. "Beach is on the
phone," she said. "He wants to talk to you. He knows, I think he
knows."

"Shit," exclaimed Kamb. "I knew Mark shouldn't have talked to
Stillman. Tell him I'm not here yet. I've got to talk to Mark." Later
that day, during a phone call with one of the editors at *Science*,
Kamb and Skolnick said they had received a worrisome call from
Beach. The editor agreed that if certain changes in the research
manuscript were made that day, publication could be set for the
next available slot, two weeks away on April 15.

Damn it, thought Beach when he hung up with Shattuck-Eidens. He called Kamb back a half hour later, but Kamb didn't return the call. He called again the next day and this time, after
Kamb took the call, Beach quickly came to the point. Again,
Beach asked, had Kamb's lab found that the melanoma susceptibility gene and his p16 gene were the same? Kamb was evasive.
Beach again pressed on, saying that if it were true, the two labs
could publish their most recent findings together. Beach was convinced now that Kamb had evidence that his p16 gene, when inherited in damaged form, put people at high risk of skin cancer.
Beach was excited since it expanded the significance of his gene.
Combined with what he and the Harris group had just turned up,

the two discoveries would make quite an impressive splash if published in tandem in a high-profile scientific journal.

Beach then asked Kamb to hold off his publication so that the two labs could release their findings simultaneously. "Geez," Kamb thought. "Mark's worries *were* justified." While he wasn't certain exactly what Beach had found out about p16, it was a good bet that his lab now had some evidence the gene played a role in causing cancer. After finally closing the conversation without making any kind of disclosure, Kamb passed along details of the phone call to Skolnick. Both men were now convinced that Beach was close behind. Maybe, they wondered, he was ahead.

Skolnick then remembered something. Weeks before he had thought about a public forum to present his finding to the scientific community and the press. He finally decided to present the findings at Cold Spring Harbor in June. But prior to that, he had considered giving a talk at a major cancer symposium in mid-April sponsored by the American Association of Cancer Research, a conference that generally received attention from the scientific and lay press. He now recalled that in browsing through the meeting's printed agenda he had seen that Beach was scheduled to make a formal presentation of his latest cell cycle findings at the meeting. He checked the AACR schedule again. There it was, still another coincidence. Beach's talk was scheduled for Wednesday, April 13, just two days before the *Science* publication date.

On Monday afternoon, April 11, Kamb drove to the Salt Lake City airport for a flight to San Francisco. Before he left, Skolnick went over his instructions. Make certain you listen to Beach's talk. Visit the pressroom at the AACR meeting. See if Beach is hanging around, talking to reporters. We don't want Beach to take the credit, Skolnick hammered over and over. We need the credit; it's our work, and we've earned it.

Eighteen months earlier, Skolnick had learned a painful lesson about pressrooms at scientific meetings attended by the national media. He had had the good fortune of having the publication of important research coincide with a high-profile scientific meeting. That episode occurred at the annual meeting of the American Society of Human Genetics. Skolnick reported the mapping of the melanoma gene to chromosome 9, which he expected to publish soon. Among the authors of the forthcoming paper was Jane Fountain, a young molecular biologist at MIT. Worried the Utah researchers wouldn't give her contribution its proper credit, Fountain visited the pressroom the day before the gene mapping paper was to be released, telling reporters that her lab had provided Utah with DNA markers that made the Utah finding possible. Several news reports the next day made highly complimentary references to Fountain, much to Skolnick's chagrin. "She staked out the room and basically told reporters that the mapping was her work, and that our contribution was minor," Skolnick recalled later. "It was amazing."

At 7:30 Wednesday morning, April 13, Patricia Molino, a veteran public relations specialist, was preparing to leave her hotel room in San Francisco when she received an unexpected phone call from her assistant, Barbara Goldberg. Molino and Goldberg were running the pressroom at the AACR. Their job was to keep the journalists covering the meeting happy, making certain they had all the free food they needed, free phone lines, and copies of research papers being presented at the five-day scientific conference.

Molino's staff had spent weeks prior to the meeting wading through the hundreds of scientific presentations for choice morsels of science to feed the reporters in attendance. On the day a research paper of potential interest was presented, the publicists worked the pressroom, talking up the science, pointing out the small hill of press releases on a centrally located table, arranging interviews with researchers and often setting up press briefings—miniature press conferences in which a group of scientists discussed related bits of scientific advances and made themselves available to questions from the press. It wasn't easy work. Much of the research was esoteric, often even beyond the grasp of Molino and her associates. When Molino could help turn science into understandable nuggets of news the press, the researchers, and the public all benefitted. But, as the events of that day were to reveal, the cancer gene research story had now become headline news, and everyone who could benefit from the story was seeking a piece of the action.

For Wednesday, the last day of meeting, Molino's team had little left to pitch. She was pleased to hear Goldberg on the phone and what she had to say. "I think there could be a lot of interest in David Beach's presentation this morning," Goldberg said, after exchanging the briefest of morning pleasantries. Goldberg was already at the second-floor pressroom of the Moscone Convention Center and had been advised that Beach "had some very interesting work to discuss."[9] According to rumors quickly spreading among the scientists at the meeting, Beach's gene was a new tumor suppressor, Goldberg said. Should we schedule a briefing for the reporters in the early afternoon? she asked her boss.

Molino knew Beach slightly as a leading explorer in an especially complicated area of cell biology. Would reporters really be interested in the intricate interplay between cyclin D and cyclin-dependent kinase inhibitors, Molino wondered? Her skepticism faded when she stepped into the pressroom a little while later. There was an unmistakable buzz of breaking news. Up on a bulletin board where the AACR press staff posted the day's schedule there now hung a note alerting reporters that a press briefing at 12 noon featuring David Beach was being inserted into the day's lineup. To experienced reporters it was a tip-off that something was up. To make certain no reporter missed out, Molino's crew decided not to be subtle: the announced press conference included the words, "hot new data," a phrase that Molino's staff hadn't used all week.

One veteran science reporter also had given one of Molino's associates a copy of an article that was to appear in Friday's issue of *Science* with Kamb's discovery. *Science* apparently also was taking little chance that the national media would miss the scientific scoop contained in its pages. It had assigned one of its own news reporters to write a story under an emblazoned headline, "New Tumor Suppressor May Rival p53."[10] Another Molino associate quickly made copies of the news article and stacked the copies on the table set aside for press releases, underlining the special text on the articles warning that the Kamb report and the accompanying news article were "embargoed"—that is, blocked from public release—by *Science* until Friday.

Distributing copies of *Science*'s news article had an unintentional but unmistakable effect. Anyone entering the pressroom that morning could now read about the momentous discovery, even though it was supposed to be kept confidential for two more days. A piece of highly privileged science news, which Kamb and Skolnick had worked so hard to control, had just been released for all the world to see. It looked as if Skolnick's paranoia was justified.[11]

The news story, written by Jean Marx, *Science*'s chief molecular biology reporter, was a red flag for reporters at the meeting looking for news. Reporters who cover biomedicine regularly monitor what Marx is writing each week to gauge what the editors at *Science* believe is the week's most important research article in the red-hot field of genetics.

And in case there was any doubt about the validity or value of the research from an otherwise unknown investigator in Utah, Marx described Kamb's discovery as "a major new addition to the list of tumor suppressors." In a quote that jumped off the page, Bert Vogelstein hailed the Kamb find as being of "phenomenal importance."[12] Marx also made very clear that the gene had been first isolated by Beach. And, in what certainly caught the eye of reporters at the meeting who saw the article, Marx reported that Beach, and collaborators at the National Cancer Institute, "confirm that they, too, have similar results" to Kamb's. Marx added, however, in a phrase once more certain to raise the competitive interest of science journalists, that Beach and the NCI researchers "are not yet willing to discuss their work in detail because it is still unpublished."

Several reporters in the pressroom guessed Beach might use his previously scheduled talk or the ad hoc press briefing at noon to discuss his own unpublished experiments showing that the p16 gene was a new tumor suppressor of note. Indeed, several reporters, increasingly cynical in the face of growing competition among scientists, were certain Beach was behind the unexpected press briefing. "It was an ideal forum for him to scoop the Utah folks," said one reporter. "No one would criticize him for releasing unpublished data at a meeting where scientists are encouraged to do that." The reporters in the newsroom figured that if Beach allowed just enough to slip out about his "similar results," they could use it as a hook for a major news story, thereby beating *Science* into print by one day and scooping unwitting medical reporters back home who were unaware of what was taking place.

Beach was oblivious to the press activity, but he did know, by the time he was about to give his talk, that Kamb's research was coming out in a few days in *Science*. He knew because Jean Marx had called him to help prepare her article, and because several news reporters had also phoned him before he left the East Coast, asking him to comment for news reports they were preparing to file on Friday, when the *Science* embargo lifted.[13]

In fact, it was during the call from Marx that Beach finally learned the true details of Kamb's research. Kamb wasn't reporting isolation of the melanoma gene as Beach had expected. And he wasn't reporting that the melanoma gene was p16. Kamb was reporting essentially the same thing that Beach and Harris had found in recent weeks and hadn't yet published themselves. They'd been beaten! Beach was angry. He passed along the news to his lab, and tried to be philosophical, noting that the field of molecular biology had become increasingly competitive and that fields of inquiry once distinct were overlapping. Nobody had expected a lab specializing in human disease to make a major find regarding the cell cycle. "These days," he later told his lab, "you just never know who your competition is going to be."

Despite telling Marx in general terms that his lab had accomplished the same thing as Kamb, Beach and Harris decided that Beach should discuss only the Cold Spring Harbor lab's work picking apart the interplay of proteins in the cell cycle. Nonetheless, just before he went up the podium to give his talk, one of Molino's associates asked him to come to the scheduled press briefing because "some reporters wanted to chat with him about his research." Beach wasn't certain what the press's interest was, but he decided he had better find out.

Meanwhile, as Skolnick had advised, Kamb was nosing around the pressroom, as he had been since late Monday, poking his head inside, or checking out the table where press releases and research papers were piled up. On Wednesday morning before walking over to listen to Beach's formal presentation, Kamb dropped by the pressroom one more time.

When Kamb saw the press briefing announcement he became concerned. Kamb walked over to Molino and introduced himself as the author of the embargoed *Science* paper. "I'm worried this is going to compromise my position," he told her. "I want to come to the press conference. I want to hear what Beach says. I really think I ought to be there."

Molino thought Kamb was pleasant enough; a young and sincere scientist. But she thought his concern was misguided, perhaps paranoid. She told Kamb only journalists or presenting scientists were invited. Kamb insisted he should be present and Molino, worried by his insistent tone that he might cause a problem if barred from the briefing, said she would look into it.

Satisfied he'd made his point, Kamb went off to hear Beach's formal talk. During the 20-minute presentation, Beach spoke at length about p16 and p21, and their roles in the cell cycle. Only in passing did he mention there was some evidence the p16 gene might be mutated in cancer tumors. But he steered away from the breaking news. Kamb was confused. Maybe, he thought, Beach already has a paper in press and doesn't want to jeopardize it by releasing details that are embargoed in another journal. But then, how did he explain the phrase "hot news" being touted to reporters in the pressroom?

Less than two hours later, Kamb was shaking Beach's hand—it was their only friendly moment of the day. Molino had agreed to Kamb's request and even suggested she introduce the two men prior to the briefing. Beach assured Kamb he had no intention of saying anything jeopardizing the embargo. Kamb then settled into a seat at the back of the press briefing room. Beach was uncertain exactly what was the proper thing to do. He hadn't read Marx's article and he didn't know it had been widely distributed among the press waiting for him. In fact, most reporters were interested in talking to Beach only because of what they had read in *Science*. They already knew his lab deserved credit for unearthing the gene and that it had performed the same experiments being reported in Friday's *Science*. Entering the pressroom Beach wondered how he could make certain reporters at the briefing knew his contribution was the foundation upon which Kamb and Skolnick had merely added a few extra rooms, and how he could do this without breaking the embargo.

Prior to the press briefing Beach called Curt Harris, his collaborator at NCI. Tall, thick-chested, with wide shoulders and a trim white beard, the 50-year-old Kansas native was widely respected by his peers. His habit of speaking slowly and in a deeply resonant voice added to his stature. Harris, like Beach, was distressed when he heard Kamb had beaten them. Harris's lab had run full-tilt for weeks, scouring tumor cell samples for mutations in p16. Like Beach, Harris believed the p16 gene's importance had yet to be seen by other labs. News of Kamb's report, passed along to him only days before by Marx and then again by other reporters, had stunned him.

But Harris knew there was virtue in accepting defeat gracefully. Doing otherwise would not be prudent, he advised. Beach decided to go to the press briefing but say nothing about their own work. He wasn't prepared for the aggressive quizzing reporters gave him. "When I got [to the briefing room] I suddenly realized that this wasn't just a 'chat' with a few reporters," he later said. "It was the real thing."

The briefing room was packed; there were klieg lights and television cameras. The press urged him to confirm details contained in the Marx article. Thus goaded, Beach reluctantly began talking about Kamb's work. Kamb then jumped up from his seat, introduced himself, and told the gathering, "Listen, this is my work. If you have questions about it, ask me." At which point, several reporters, surprised and excited by Kamb's unexpected appearance, shouted that he should come up to the podium. Beach sat down.

Watching all this unfold, one reporter recalls thinking that to adhere to the embargo was like trying to defy gravity. It wasn't going to happen. Even Kamb realized that. Up at the microphone he answered detailed questions about the importance of the new cancer-causing gene. He reiterated that he thought it might be mutated in many cancers, since his lab had found alterations in the gene in almost half of all tumor cell lines they sampled. A reporter asked if speaking out in public didn't break the embargo. Can't we report now what you've been saying? Kamb conceded he thought the embargo was now a "farce." He told the reporters he needed to make a phone call and he'd let them know shortly, through Molino, if the embargo was no longer in effect. Sitting nearby, Molino was stupefied. Never had she seen someone actually succeed in organizing his own impromptu press conference in the middle of one set up for someone else.

Kamb then called Daniel Koshland, the editor of *Science*, who wasn't surprised to hear from him. It wasn't the first time the two men had spoken. Several days prior to leaving for San Francisco, Kamb and Koshland talked about what they should do if Beach discussed the Utah finding during his formal presentation. Koshland, himself a molecular biologist with a lab at the University of California, Berkeley, liked the younger scientist. Kamb had worked in the laboratory of one of Koshland's close colleagues; Koshland even had recommended Kamb for a job. He was impressed by Kamb and wanted to help him succeed. *Science* had been getting calls all day from reporters wondering whether the embargo was terminated. "I was trying my best to protect a promising young scientist," Koshland recalls thinking. "I thought [what was going on in San Francisco] was absolutely outrageous." The two men agreed to formally end the embargo.

Back in Utah the next day, Skolnick reveled in the attention. Yet, try as he might, he couldn't swallow his irritation. As he had hoped, the *Science* paper was hailed by the national media. *USA*

Today, on its front page, carried a story under the headline, "Missing Gene Linked to Many Cancers." Similar stories appeared in *The Wall Street Journal*, *The New York Times*, the *San Francisco Chronicle*, *Newsday*, and the *Boston Globe*. Many other publications, including *The Washington Post*, ran a story filed by Richard Cole of the Associated Press. The *Orange County Register* put Cole's story on the front page, headlined "Key Cancer Gene Is Identified."

In almost all the stories, Myriad Genetics was mentioned prominently. But Skolnick was miffed because every story gave almost equal credit to Beach, something Skolnick was certain was a result of Beach's active presence in San Francisco. Moreover, several stories, including Cole's AP story, made it seem the discovery was being released at the AACR meeting in San Francisco, and not in Myriad's research paper in *Science*.

But then, four days later, *Time* magazine ran a cover story on cancer genes in which Skolnick, Kamb, and Myriad played a central role. *Time*'s science correspondent had been working on a story about cancer gene research, and the *Science* paper was the news hook needed to get the article published. On the article's third page, following a huge picture of a malignant cell, was a black and white photograph of Kamb sitting next to his lab bench. Under the photo, *Time*'s editors said, "The multiple tumor suppressor gene that Alexander Kamb and his colleagues discovered may explain why . . . melanoma cells went astray." *Time* downplayed Beach's role, noting only that his lab had previously discovered the protein produced by MTS1, and adding, "last week [in San Francisco] it became clear that p16 and MTS1 are one and the same."

Kamb, of course, was pleased. His family telephoned with congratulations. Even his grandfather, Linus Pauling, called with warm wishes. "That really made me feel proud," Sasha told his friends, noting that to date he didn't think his grandfather had thought much of what he was doing. Vogelstein sent a congratulatory telegram, praising the discovery as "the most important of the year." And then, topping things off, the following week, in *Sci-*

ence's chief competitor, the British journal *Nature*, researchers at the University of California, San Diego, reported exactly the same thing.[14] Not only did the *Nature* article confirm their findings, but it convinced Kamb and Skolnick they were right to rush the results into print. "We were very lucky, very, very lucky," Skolnick recalls.

13: The Roller Coaster

By late March 1994, Donna Shattuck-Eidens was getting angry. Three and half years after Mary-Claire King uncovered the existence of the breast cancer-causing gene and mapped it to somewhere in the middle of chromosome 17, Shattuck-Eidens believed that her gene-hunting crew at Myriad Genetics was so close to the gene's precise location that "I could feel its presence," she says.

Shattuck-Eidens is upbeat and confident, and her short strawberry-blonde hair, freckles, and self-deprecating giggle can even make her seem somewhat perky, not the typical image of a hard-driving, gene-hunting molecular biologist. For the previous six weeks, Mark Skolnick and Sasha Kamb, her bosses at Myriad, had temporarily pulled her team off the BRCA1 search, ordering her and her lab workers to put all their time and resources into the race to identify MTS1's role as a major cancer-causing gene.

But with that work nearly complete, Shattuck-Eidens worried that her staff had lost the focus she had instilled in them during a relentless pursuit of BRCA1 during the fall and into the winter. While many of her friends and family spent their weekends—and even many weekday evenings—on the nearby spectacular ski slopes in the Wasatch Range, just 45 minutes away from Myriad's labs, Shattuck-Eidens was glued to her desk in her small office, reading print-outs and x-rays produced by her lab workers, as they carefully eliminated, one by one, 40 different bits of DNA from a tiny section of chromosome 17 where, she was certain, BRCA1 lay hidden.

But having abandoned BRCA1 for more than a month, she was anxious and getting mad. Almost every week her e-mail bristled with rumors that one of the half dozen other teams of molecular biologists from the United States, Canada, Great Britain, or France already had beaten her to the breast cancer gene. In early March, a bit of scuttlebutt buzzed about the Internet hinting that Francis Collins was preparing to proclaim that his collaboration with Mary-Claire King had finally won the race. But, it turned out, the hubbub was simply over a press conference Collins was hosting at the National Institutes of Health to announce that Bert Vogelstein and Richard Kolodner had crossed the finish line neck-and-neck in their race to find the second HNPCC (colon cancer) gene discovered in less than three months. A week later, a researcher at the University of Utah working with Myriad rushed over to say that that he'd heard from a reliable source that Donnie Black, a biologist they all knew in Scotland, had found BRCA1. Referred to as the "Black rumor," it set Skolnick, Kamb, and Shattuck-Eidens sweating for hours until they tracked down the source of the story and found it was erroneous.

Skolnick had hired Shattuck-Eidens in 1993 to analyze lengths of previously unidentified DNA that were fished from the genetic material of individuals in breast cancer families. Her job was to determine whether any of the material extracted from someone with cancer contained a mutation. The researchers assumed that finding a mutation in the length of DNA being analyzed was most likely occurring inside the confines of a gene that caused cancer. Once the mutation was found, it would be relatively simple to isolate the surrounding gene.

Indeed, in the previous October, not long after she joined Myriad, Shattuck-Eidens persuaded herself that she had fallen upon such a mutation in one family member's DNA. The lab had found a series of nucleotide base letters that were out of place, suggesting a defect that signaled the presence of a mutated gene. For days on end, her lab compared the DNA anomaly to the sequence in the same area in other family members. But after three weeks, including one "heart-stopping weekend," she says, the "mutation" turned out to be simply a stretch of nucleotide units that often varies, quite harmlessly, from person to person.

"You know, what we do, screening for mutations, is very tedious, very boring work," Shattuck-Eidens says. "It's repetitive, doing the same experiments, one after another. When we get a 'gene scare'—that's what we call it when we go a few days or weeks thinking we've latched onto the gene—then everyone gets revved up and works around the clock. That's exciting. Of course, each time it turns out to be a false alarm. Most of the time, the work is very stressful."

Shattuck-Eidens, however, was new to the race, energized, optimistic, and most important, one of the few fresh players on teams that were beginning to grow deeply weary of the game. Before Skolnick hired her, she was unhappily working at a plant genetics company that was losing money. Suddenly, she was catapulted into one of the most widely watched gene hunts ever, competing against some of the world's foremost scientists, including the now-legendary Mary-Claire King.

But in Berkeley, King was becoming exhausted by the effort. By 1994, she had been seeking the gene on and off for almost twenty years. Compared to others in the field, her lab was understaffed and underfinanced, and tensions were rising between her group and her collaborators. Sometimes, King sighed (and told a visitor to her lab in March of 1994), it was difficult to remember how far she had come. "It's been such a long effort," says King. "Sometimes I'm just not sure how much longer I can keep it up."

Throughout the early 1980s, King continued to collect families, hoping that a large sample of big families affected by disease would provide the kind of statistical power she needed to find a link between some type of genetic marker and the gene itself. As a result of her many published research papers and the scientific talks she gave around the United States, her interest in analyzing the genetics of breast cancer soon was widely known in the medical community. Doctors and other public health officials she ran into often passed along the names of patients who reported multiple instances of breast cancer in their families. Often, aided by a graduate student or two, it took her months, sometimes more than a year, to track down the affected members of a particular family and determine if the clustering appeared inherited or simply a case of familial bad luck. It was slow, laborious work; aside from Henry Lynch in Nebraska, who conducted his inquiries through word of mouth and by querying patients he treated, and Mark Skolnick in Utah, who was scanning the Mormon genealogies, nobody else was conducting a formal effort to compile families in whom a gene appeared to be at work.

One resource that proved especially valuable to King was a landmark government-funded study performed between 1980 and 1982 that identified more than 1,500 women who had developed breast cancer before they reached age 55 and had had their cancer treated in one of several medical centers in and around Detroit and San Francisco. Within six months of a diagnosis of disease, these women were interviewed at their homes by researchers working under a grant from the National Cancer Institute. The women were surveyed about their diet and their environment; and at King's request, the women were asked whether there was a family history of their disease. When the study was reported, King asked the investigators for permission to contact women who had said their mother or sister also had breast cancer.

As a result of her own inquiries, by the middle of the 1980s King and her colleagues identified a few dozen especially interesting families. Through months of phone calls and personal interviews, King uncovered one family in particular in which a 36-year-old woman with cancer reported that she had a sister, a paternal aunt and, surprisingly, a paternal uncle with breast cancer, too. Over the course of more than a year, King tracked down 77 members of the original woman's extended family. By reviewing medical records and death certificates, researchers working with King eventually found 14 confirmed cases of breast cancer, 11 in women and 3 in men, among 252 adult relatives spanning five generations.

But it wasn't until 1986, with the arrival of two young researchers in her lab, that King was finally able to muster the emerging power of the gene-hunting technology that was sweeping through molecular biology labs in the United States and abroad. Jeff Hall, a young biochemist. Within a few months of his arrival, Hall completely overhauled the kind of science taking place inside the King lab, using new DNA-based markers to scan genetic material extracted from blood of families being retrieved by King. The other arrival was Beth Newman, a doctoral candidate in epidemiology who, one day, made an extraordinary insight into King's family data. Together, Hall and Newman provided ways of uncovering clues about the gene that until then had evaded King's grasp.

King hired Hall to scan the entire vast genome looking for DNA markers that might be linked to a putative breast cancer gene. By 1986, scientists using RFLP markers had stunned the world of human genetics by mapping the general location of the genes that scientists had been struggling to uncover without much success for years. In 1983, researchers pinpointed the approximate site of a gene for Huntington's disease, and by 1986 other RFLP-toting scientists were closing in on the genes underlying cystic fibrosis, muscular dystrophy, and several other inherited diseases known to be caused by a defect in a single gene. "You have to hand it to Mary-Claire's persuasive grant-writing skills," says Hall. "It took a big leap of faith to argue you could use the new markers to find a gene for a disease that many people still argued wasn't even inherited."

But Hall was upset with King's lab, with some of its small staff untrained in DNA techniques, the lack of molecular biology equipment, and the family material he was told to analyze. Many blood samples from families were old and useless. "We realized we'd have to draw blood [for new DNA samples] from many of the families, or we'd need new families, too," Hall says.

If several unexpected events hadn't transpired, Hall might still be mucking around for the gene's existence. Perhaps the most important development had occurred in 1985 when a California researcher discovered a breakthrough tool for molecular biologists that he called polymerase chain reaction, or PCR. With it, scientists like Hall can take a tiny bit of DNA and amplify it many times over. PCR breathed new life into many of King's old blood samples, and allowed Hall to do many more experiments without waiting for blood to be drawn from the family relatives. At about the same time, Ray White and others found a new class of RFLPs that were much more commonly sprinkled about the chromosomes, giving Hall many more targets all over the genome.

Most surprising, however, was some unwitting aid King received in late 1987 from First Lady Nancy Reagan. A television crew from the ABC-TV affiliate in San Francisco had filmed an interview with King in which she described her search for cancer-prone families. The next morning, Mrs. Reagan was diagnosed with breast cancer. Seeking background material on research into breast cancer, 127 affiliates of ABC-TV across the United States ran the King interview that night.

"Within a few days, we heard from 3,000 women," King recalls. Soon she and her research assistants had identified about 100 families with many cases of breast cancer. Among those calling in the days after the television broadcast was a 65-year-old woman from Illinois named Betty. She called because her 71-year-old sister had just been diagnosed with cancer, and she was deeply troubled by what she heard King say on TV. Betty herself had had breast cancer at 45 and ovarian cancer at 61. Although two of Betty's other sisters also had had cancer years before, and so did several of her siblings' children, it wasn't until she heard King talk about a possible genetic link that Betty realized the disease in her family might be inherited and that her own 33-year-old daughter was at risk.

King had conflicting reactions to calls such as Betty's. "As a scientist I was overjoyed by learning about a family like this, a family that made the case for a gene very strong," King says. "But it was very upsetting for us, too. This woman likely had a medical condition in her family that she either wasn't aware of or had been able to deny. But we couldn't tell her she had a gene for certain, nor could we tell her if her children were at risk. Still, she and others joined the research because they wanted the answers even more fervently than we did."

Betty's family, formally known as "Family One" (see accompanying pedigree) in King's subsequent research reports, was one of twenty-three large families to surface during this time. But even with the larger families, the large number of markers, and the high-quality samples, Hall wasn't making much headway, and he was getting frustrated.

Hall also sometimes complained to King that she was diluting the lab's focus by taking on too many projects for their small operation, such as her efforts to use the power of genetics to help Argentine grandmothers locate children taken from their parents during that country's repressive regime in the 1970s. Moreover, she briefly joined the effort to use the new RFLP marker technology to find a gene for cystic fibrosis, even though that gene hunt had attracted some of molecular biology's hottest practitioners. She also began exploring the genetics of lupus and also began a project involving AIDS.

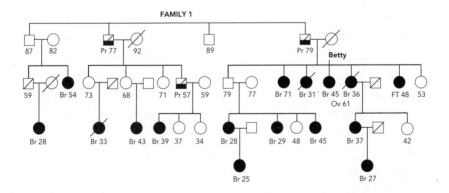

The above illustration is a pedigree of a family known as "Family One" in Mary-Claire King's research reports. Black circles represent cancer and age of onset; diagonal lines represent gene carriers.

"It was all a big distraction," says Hall. "I kept saying to her, 'We have this grant for breast cancer. This is where we'll make our fame and fortune.' In order to be heard I had to yell, 'Hey, let's get this breast cancer thing really going!'"

Hall, who later took a job with a biotechnology company in Seattle, says he has "tremendous respect" for King. "Mary-Claire is curious to a fault," says Hall. "That's also her strength. I admire her greatly. But I gotta tell you, if we'd gotten scooped I guess I would have complained forever that we should have just looked at breast cancer."

For her part, King concedes she was still feeling her way with the new DNA probe technology. Like others in human genetics she was admittedly awed by the power of the new weapons against diseases that for decades seemed beyond the rational dissection of modern science. Although she acknowledges this only to a few close friends, she also wasn't confident that "the breast cancer project" would really pan out. Putting all her laboratory eggs in one scientific basket was too big a bet to make.

But Hall, 33 years old and itching to build his career, believed he had nothing to lose. Between the spring of 1988 and the summer of 1990, "the lab was cooking," he says. All told he fired over 180 different markers of various types at chromosomes all up and down the genome.

Testing each of those markers took Hall and the lab technicians working with him a few weeks. And, more troubling, he wasn't convinced that the families he was testing were all linked to the same gene, or even to one gene. "Finally, Beth [Newman] came up with a solution," he says. Newman was helping King gather and analyze data from the large pedigrees. One day she suggested that King's group arrange the pedigrees in an order based on the average age of onset of breast cancer in each family. Once the lab did that, a consistent and strong pattern of inheritance arose among the families in which women were developing cancer sometime between the ages of 45 and 50. King, Hall, and Newman decided they would have the greatest chance of finding a gene by restricting their marker search within the DNA from those so-called "early-onset" cancer families.

"It was a very significant breakthrough for us," King says. "We now had a powerful model of inheritance; all we needed was a linkage marker."

Soon afterwards, Hall found a specific marker on chromosome 17, called D17S74, that was consistently linked in seven of the twenty-three families under intensive study. The seven families were precisely those that Beth Newman had ranked as having the largest clustering of cancer in women between forty-five and fifty years old. If a defective gene were inherited, it was likely to trigger cancers earlier in life, whereas sporadic cases of cancer—those not related to an inborn defect—would generally be expected to take longer to develop.

"We got very excited, but Mary-Claire had been burned before and she wanted to be certain," says Hall. "All of this is so much luck and perseverance; I don't think we were prepared to believe it yet." Hall tested several other markers nearby and produced similar statistical results, showing in the seven families that markers were present consistently in people with the disease, and just about never in those who were disease-free. This meant that the marker and the alleged gene were located so near to one another on chromosome 17 that they were consistently being inherited together.

Still, when researchers took a look at King's results after she announced them in Cincinnati in October 1990, it was determined that the gene and the marker were likely to be about 20 million base pairs apart. Clearly, the most important aspect of her report, of course, was proving that a gene existed at all. But the marker also pointed to the region in DNA where the gene was located. Considering that a full complement of human genes is three billion base pairs long and a typical gene is about 20,000 to 30,000 base pairs long, the "narrowed" region on the long arm of chromosome 17 identified by the King lab probably contained some 300 to 500 genes, each of which could possibly be the one that causes inherited breast cancer.

"It really made sense to hook up with Collins," Hall says. "Doing linkage analysis is one thing, and no small feat at that. But the next step required us to do something else again." Indeed, the job ahead involved plucking out of the narrowed region bits of DNA thought to be part of suspected genes, expressing as much of the gene as possible, and then searching the length of the gene for a mutation, for a bit of the DNA missing or out of place.

Collins was one of the masters of a new gene-sleuthing strategy called positional cloning, a set of molecular biology techniques and tools that, though tedious, time-consuming, and filled with opportunities to make very big mistakes, allowed a persistent and lucky scientist to draw a bead on a hidden gene within a specified region such as had been identified by Hall and King. "It's not a perfect science; maybe it's more of an art," says Collins, a supremely confident and, some would say, even cocky scientist. ("I'd rather you said I'm willing to take some chances," he says.)

Unlike King, Collins had the resources to juggle several scientific balls at once; he relished being a contestant in numerous high-profile scientific events; he was becoming America's iron man of molecular biology. Working with King, he expected the DNA from the "narrowed region" through the rigors of positional cloning, devising new techniques as needed along the way. Understandably, King was overwhelmed by Collins's apparent mastery of positional cloning. "We hoped and expected that [Collins] would help us out, bring us along, that bringing our lab up to speed was part of the bargain," King says.

In positional cloning, scientists take a bit of DNA known to be within a region and use it somewhat like a bit of adhesive tape to snag larger pieces of genetic material in the same vicinity. One of the best ways to snare a large segment of DNA was to grow swatches of material in yeast cells, so called yeast artificial chromosomes (YACs), that could hold as many as one million base pairs of DNA at once. By overlapping the YACs one atop the other the researchers could slowly build a "contiguous" physical map of the DNA within a specific region. The method was much like drawing a detailed map of the geography along a highway by laying successive one-mile lengths of road over one another, superimposing the mile-long lengths at half-mile intervals. The

DNA cartographers knew each successive bit of DNA was connected to another because the overlapping part of the genetic material was identical with the stretch of DNA below it (see
accompanying drawing).

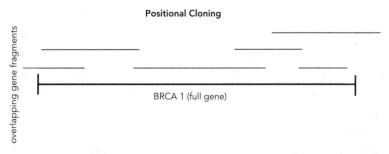

The figure above is a rough representation of how a full-length gene is
pieced together with overlapping gene fragments.

But before they could conduct this leapfrogging effort down the
length of the chromosome, the researchers needed to significantly
narrow the region of their hunt. Toward this end, King decided to
follow two broad approaches, both of which required plenty of luck
and exploited well-known quirks in DNA replication (when a cell divides and must duplicate its full complement of genetic material).

Genes sometimes have a way of becoming physically unlinked,
that is separating from one another, as they pass from generation
to generation. This occurs when a new germ cell—an egg in the
female, a sperm in the male—is being formed. Unlike the rest of
the body's cells, a germ cell has one-half of each of the twenty-
three pairs of chromosomes. That's because at fertilization the
sperm and egg each contribute their one-half to make a complete
set of chromosomes in the new offspring. When the new germ cell
is being formed, it selects pieces from the original set of paired
chromosomes, somewhat like a diner in a Chinese restaurant picking some dishes from column A and other dishes from column B.
The rules the germ cells follow in this process of recombination, as
it's called, are unknown, so recombination is considered unpredictable. It never occurs exactly the same way twice, which ex-

plains why siblings are different from each other even though they inherit their genes from the same two parents.

Exploiting natural recombination events is a valuable gene-hunting technique. As a result of recombination, a genetic marker that appears linked to an unknown disease gene in one or two generations can become unlinked in some members of later generations. The gene and the marker separate at some point during recombination. But the closer a marker is to a gene, the greater chance it will survive recombination and be inherited with the gene through the generations. (See accompanying figure.)

Since recombination events in a specific region can't be predicted, King's tremendous contribution was her extensive families, in which numerous opportunities existed to find recombinant events. In the figure of King's research "Family 7," a particular recombination helped scientists move their search up along a chromosome. Parent B, the mother in this pedigree, likely carries the gene because she has cancer. Markers tracing down the length of chromosome 17 show that children 2, 3, and 5 who also were affected inherited Parent B's left-side chromosome, the likely site for the mutated gene. Daughter 4 isn't affected because she inherited the right-side chromosome.

As can be clearly seen, daughter 1 still developed cancer even though, through a recombinant event, she didn't receive the bottom three markers on the left-side chromosome. And daughter 4 was unaffected even though she did receive the bottom two markers. Thus, by studying this family with numerous markers close together, the researchers realized that the gene had to lie above marker "NM23." This family's recombinant event helped shrink the search area.

But finding recombinants is a "dreadfully slow process," says King. "And it's why we couldn't stop searching for more families, and more people in the families we had. Finding a recombinant is luck and a lot of looking."

So, at the same time, King decided to pursue a long shot she hoped would catapult the researchers to the gene. It led her to one

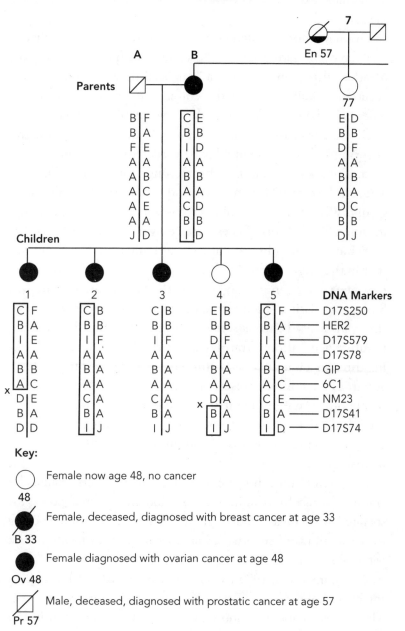

FAMILY 7

Key:

○ 48 — Female now age 48, no cancer

● B 33 — Female, deceased, diagnosed with breast cancer at age 33

● Ov 48 — Female diagnosed with ovarian cancer at age 48

◻ Pr 57 — Male, deceased, diagnosed with prostatic cancer at age 57

The figure above represents how a chromosome recombination allows scientists to narrow their search for the location of a gene by using DNA markers.

of the most exciting few weeks in the four-year effort to identify the gene.

This new effort entailed making use of still another peculiarity that can occur during germ cell division. Much rarer than recombination, this event is called a "balanced translocation." In this occurrence, half of one of the chromosome sets fuses with the other half of the set, like square dancers switching partners.

Often this crossover doesn't produce any problems for the offspring as long as the fused parts carry their genes intact with them. But sometimes as the chromosomes join, a bit of DNA or a gene is lost, resulting in some kind of abnormality in the offspring. What King was looking for was a child born in a family in which fused chromosomes were being passed along mostly intact, but in whom breast cancer was common. The hope was that in the original translocated chromosome that first occurred in some earlier generation, a bit of the breast cancer gene was broken or missing. An affected person whose cancer was due to this translocated chromosome would pass the gene defect along to future generations. Translocations can be visualized by looking through a microscope at chromosomes cultured from a fetus as part of an amniocentesis study. Seeing the exact spot of the breakpoint can help scientists home in on the gene being pursued. Indeed, Collins was much praised for using the balanced translocation technique as a shortcut to finding the neurofibromatosis gene.

King sent out a call to numerous medical centers around the country for records of amniocentesis, looking for breakpoints associated with breast cancer in the family. But despite poring over thousands of reports, no breast cancer–translocation link was uncovered.

By the middle of 1993 the fervid race for BRCA1 had gained international media attention. King was under intense pressure. Through recombinants and by finding new markers within the region, the area under study had been reduced to one that was still large enough to contain about 24 genes.

King, though, was beginning to despair. Over the previous two years, her team had tried every shortcut technique known. One that was particularly frustrating involved looking at genes in the chromosome 17 region that had a good likelihood of being involved in breast cancer. Indeed, soon after King and Collins first joined forces, King had come across several intriguing genes that were known to reside in the region that, she was certain, were actually BRCA1.

For many months King's lab tested for mutations in a gene called EDH17B, that was known to interact with estrogen. To King this was an especially attractive candidate because scientists know that the female hormone estrogen promotes cell growth. Too much estrogen might cause cancer. The EDH17B gene produced protein that sits on cell surfaces and grabs hold of estrogen, drawing it into the cell. Perhaps when mutated the protein absorbs too much estrogen, or somehow takes in estrogen when it shouldn't. But after almost a year of examining it, sequencing its nucleotide bases, and scanning for some minute mutation in women with cancer, King had had to admit defeat.

Similarly, the lab pored over other candidate genes in the same way. But one by one mutation analysis forced them to eliminate a thyroid hormone receptor gene, retinoic acid receptor gene, anti-metastasis gene called NM23 that everyone in the lab was hopeful about, and a collagen skin tissue gene. As each gene fell by the wayside, "we began to feel like we were on an emotional roller coaster," King says. "We go up with expectations and optimism and then each time we've come down, hard. We've had to learn to be very professional, to live with some very deep disappointments and move on."

But, as King and Collins grew weary and trudged ahead, largely unknown to them, Mark Skolnick's team in Salt Lake City was growing strong and vibrant. Ever since Skolnick was introduced to the power of the Utah genealogies and the state's extensive cancer registry, he knew that linking the two would reveal genetic forces at work in cancer, most especially colon and breast cancer, two of the most common forms of the deadly disease. But, like Mary-Claire King, Skolnick's strength was analyzing data and setting strategies. He wasn't a gene jockey. He also lacked the resources to fuel a major assault on the genes, even though the families he and his colleagues were collecting were probably among the richest pedigrees anyone was gathering anywhere in the world.

"It just drove me crazy," Skolnick says of his inability to win grants large enough to hire the molecular biologists he needed. By 1988, Skolnick realized that National Institutes of Health funding would never provide the financial resources to compete against Collins or Vogelstein. Other academic researchers were turning to the private sector to finance large projects, and Skolnick decided to "look around," he says. "Corporate research was still seen as somewhat unseemly. But I had this great idea—to find the genes of common diseases—and nowhere else to turn."

In early 1991, plans for Myriad Genetics finally fell into place. For months, Skolnick had been talking up his idea for a company with Peter Meldrum, a local businessman who had previously run an agricultural gene company. Skolnick sold Meldrum on his idea for a company that would clone genes for common diseases, and generate revenue by licensing development rights to the genes to pharmaceutical makers and by creating a genetic testing labora-tory to provide genetic testing to identify people susceptible to the disease-causing genes.

Later in 1991, the two men incorporated the company and created an advisory science board that eventually included Walter Gilbert, a Harvard University Nobel laureate in biology. A few weeks later, Myriad made a deal to receive an additional $2.8 million over three years from the giant drug maker Eli Lilly and Company for the rights to develop new medicines based on the breast cancer gene—if Myriad found it and won a patent for its discovery.

"All of a sudden," Skolnick says, "I go from nowhere to being very competitive. By then, I was way behind [in the hunt for BRCA1]. But we were quickly able to put together a very large group." At the same time, Myriad secured its private financing, Skolnick's university group finally won a small NIH grant. (Like many university professors who help form companies based on their research ideas, Skolnick was able to maintain his university affiliation while employed as Myriad's director of science.)

The joint affiliation, while becoming increasingly common among biologists seeking greater resources, is still controversial. Fellow scientists, some motivated by jealousy, some by anger, argued that researchers with a corporate connection might slow scientific progress by withholding from colleagues in the field important discoveries that were deemed proprietary. Some critics also believed that building a corporation and private wealth based on university research that had been funded by government grants was simply unethical.

But almost every major biotechnology company formed in the late 1980s and the 1990s was built around technology developed by a university scientist. As a concession to the University of Utah, Myriad agreed that it would share patent rights with the university for any gene discovery it made that involved university research, which, of course, included many of the families that Skolnick had linked to various cancers.

By the spring of 1993, under the direction of Sasha Kamb, whom Skolnick had hired during the summer of 1992, Myriad's staff was turning into a well-oiled operation, carefully managed and motivated by Skolnick. "I had the kind of team I'd always dreamed of creating," he says. "I worked hard to make sure we didn't screw up."

Researchers at Skolnick's university operations, led by David Goldgar and Lisa Cannon-Albright, continued to collect new families, searching for relatives with new recombination points that would narrow the region under study. Meanwhile, other researchers at the university, another group inside Myriad, and two government scientists in North Carolina recruited into a collaboration with Myriad were fishing out new sections of DNA that had the structural appearance of genes.

These DNA fragments were then passed along to Donna Shattuck-Eidens and her team of three lab technicians, who compared the sequences of nucleotides in the new gene fragments with similar stretches of DNA in family members, trying to determine whether there were any differences in the series of DNA nucleotides between an affected and unaffected person. Any series of letters that was out of place in DNA from a person with cancer was a hint to the scientists that they were looking at a potential disease-causing mutation.

As the hunt for the gene dragged on, David Goldgar decided to try one more big push to significantly narrow the region by uncovering new recombinants. This required looking at the DNA of a family with an especially large number of people affected with disease, and then saturating the site under study with as many markers as possible. He decided to reanalyze a family first uncovered by scanning Skolnick's Utah Population Database, the computer program that linked the Utah genealogy and the state cancer registry. By looking through the genealogy for anyone with more than one cancer in a nuclear family, Goldgar found an extended family, called 2082, with four relatives with ovarian cancer and two members with breast cancer, all of whom developed the disease before they turned 45.

Goldgar and his associates worked to expand the family, searching for new members in the genealogy, interviewing relatives about their health status, inquiring about relatives who had moved away from Utah, and scanning the cancer registry for family members not included in the genealogy. Finally, Goldgar had amassed a giant family, four generations teeming with cancer; 30 individuals with breast cancer and 20 women with ovarian cancer. It was the largest family ever found that appeared to be linked to BRCA1.

By late winter of 1993, Goldgar and his associates had collected blood from 195 family members and the DNA had been analyzed by researchers at the university and at Myriad. After many months of searching, the scientists finally found a recombination in a 45-year-old woman with ovarian cancer. By April, after a concerted effort to identify markers near the recombinant, "we were able to move the search area into a really small segment of the chromosome," says Goldgar. Suddenly, the search area was narrowed from one that likely contained 24 genes to one that could only contain between 6 and 10 genes. "If David's recombinant event was right, it meant we'd all spent the previous year looking for the gene in the wrong place," says Skolnick.

At National Institute of Environmental Health Sciences in Research Triangle Park, North Carolina, Roger Wiseman and Andy Futreal, who were collaborating full-time with Skolnick's teams at Myriad and the University of Utah, began to bore in on the new smaller sector of the chromosome. "That's when I really believed we were going to find the gene," Shattuck-Eidens says. "Mark had put together a really strong team, and his skill was making certain we were all coordinating our efforts every day."

Meanwhile, the King-Collins collaboration was beginning to falter. In 1993, Collins was named the director of the National Center for Human Genome Research, succeeding James Watson, the codiscoverer of the double helix of DNA, and throughout the winter of 1993–94 he and his staff were distracted by moving his laboratories from Michigan to Bethesda, Maryland. At about the same time, Barb Weber, who had been running Collins's lab in Michigan, was recruited by the University of Pennsylvania to run her own genetics laboratory, and she spent many months moving her operation and family to Philadelphia.

In Berkeley, King's BRCA1 group—primarily three doctoral candidates and a few technicians—was affected by the turmoil. "Nothing's slowing me down," Eric Lynch one of King's researchers on the project, told a visitor. But, even so, Lynch, whom King considered a "technological whiz," was forced to split his time, spending some days trying to complete his Ph.D. dissertation, a project to identify a gene for a type of inherited deafness.

In the spring of 1994, as Myriad sharply reduced the search area, Lynch and King believed they had eliminated about 14 of the genes in the 24-gene region they were still studying. In an effort to lighten the stress in the lab, Lynch and the two others on the project, Lori Friedman and Beth Ostermeyer, began naming the separate gene fragments after characters in the various *Star Trek* television shows. Up on a large wall describing progress of the project, the researchers listed the genes with such names as Data, Worf, Quark, Diana Troi, and Dr. Crusher.

By late spring, King had the lab focus its efforts on the gene fragment that was code-named Odo. In some family members the gene appeared to have a disruption that differed from sequences seen in healthy family members, a sign of a mutation. "I was pleased to see the variants and hoped and hoped we'd finally found our gene," King said. "But we'd had false leads so many times before. I didn't dare hope too hard." Indeed, when the lab looked harder, by analyzing the same DNA segment in other families, the shifts in letters turned out to be simply benign polymorphisms, areas in the gene that normally differ from person to person without causing any health problems.

Unknown to King, the Myriad team's candidate gene was looking more and more like the real thing.

In April, Shattuck-Eidens received a bit of DNA from Wiseman and Futreal in North Carolina. At first, the fragment seemed as if it were parts of several genes. But within a few weeks it was determined that the fragment was part of another bit of gene previously fished out by one of the University of Utah researchers, Yoshio Miki. Shattuck-Eiden's researchers—Wei Ding and Tanh Tran—were analyzing lengths of DNA by running electrophoresis gels that can detect variations in gene sequences. One day in late June, Tanh found a slight variation in the gene fragment originally sent from North Carolina. Shattuck-Eidens ordered her team to focus their effort by analyzing the gene fragment in DNA samples from different affected members of Goldgar's family 2082.

Then on July 13, Shattuck-Eidens strolled over to Tanh's lab bench and found a "funny-looking gel," she says. Tanh had gone home because his wife was in labor. "I let out a scream," Shattuck-Eidens says. "It was very embarrassing. I kept saying, 'This is BRCA1! This is BRCA1!'" The gel clearly showed a mutation in a segment of the gene that, based on the pattern of base pairs, was likely to be a so-called coding region, a series of base pairs that is crucial to the production of the gene's protein. "A mutation in that area is a dead giveaway," Shattuck-Eidens says. "When I asked Thanh later why he hadn't shown it to me before, he said that he thought he had messed up the gel. He thought the mutation was simply an error in loading the gel. Of course, it was not."

Indeed, Sasha Kamb told Shattuck-Eidens to keep looking for more mutations to be certain. At that point, Goldgar and his crew were dispatched to some members of family 2082 to draw more blood for analysis. "People were working around the clock," Shattuck-Eidens says. "It was a frenzied time. There was a week I was worried about people's health. Someone had a car accident driving back to the lab from dinner late one night."

By the end of July, Myriad had found more pieces of the gene in DNA from family 2082. And by the end of August, "we knew we had the gene," says Skolnick. "People ask me how I felt. You know, we didn't pop champagne, we didn't go out to dinner. We just felt this tremendous relief. After all these years, I could finally take off my racing shoes."

Myriad rushed in a patent application and in the first days of September sent a research paper describing the discovery to *Science* magazine. But within days, rumors spread that the gene was found, and soon King and Collins and others had been able to confirm the news: Skolnick had found the gene. On September 13, NBC News chief science reporter Robert Bazell broke the story of the gene's discovery, followed the next day by an article in *The Wall Street Journal*. For several days, Bazell had called the contestants for comment. Skolnick declined to say anything even though Bazell told him he was convinced he had the story. "The paper hadn't even been formally accepted yet as far as I knew," Skolnick says. "I couldn't comment; I tried to keep it quiet."

Bazell called Collins, too, and told him he was running a report that night. Dispirited, Collins and some of his research team walked over to Collins's home near the NIH campus and watched Bazell tell the world the race was over. The next day, Mary-Claire King put up a brave front. She issued a press release congratulating Skolnick and hailing the discovery as a major step forward. She noted that the gene Skolnick had found was very large and posed quite a challenge for scientists hoping to unravel how a defective version caused cancer.

"There's a great deal more work to be done by geneticists before women have results from this gene that will matter to our lives," she said in her formal press statement. "It's been a privilege to be involved in the breast cancer story for the past twenty years. Now we begin the next twenty."

The discovery of BRCA1 is expected to illuminate finally the hidden biochemical pathway by which a breast cancer cell becomes cancerous. Indeed, two years after it was first identified, researchers began developing evidence that BRCA1 may be a type of DNA repair gene, much like the genes found to cause inherited colon cancer, a clue certain to lead eventually to new and powerful drugs.

But, if history is a lesson, turning the discovery of BRCA1 and its mechanism of action into future treatments may in fact take, as Mary-Claire King suggested, several decades of hard work. That, at least, is what one pioneering drug company researcher in a small town in Pennsylvania found when, in the late 1970s, he launched an assault on one of the very first cancer genes discovered, with the strange name of "ras."

14: Ras

Thirty-five miles northwest of Philadelphia, the four-lane Sumneytown Pike slices past cornfields, single family homes, condominiums, and occasional retail strips that speckle the gentling rolling hills. It is here that the road intersects with the thinly populated exurb of West Point, Pennsylvania, a village so small it doesn't appear on most state road maps, and then passes by an understated corporate campus squatting off to the south. Only an unadorned sign planted out front with the name *Merck* alerts the curious motorist to the compound's identity.

Through the security guardhouse and onto the winding road-ways, past expansive parking lots and manicured lawns, there is nothing to advertise that these buildings sprawling over more than 300 acres house the world's most successful pharmaceutical laboratories. Here, beginning in 1982, a small team of scientists quietly set off in search of a cure for cancer.

When biologist Edward M. Scolnick arrived here that year and began assembling that team he was not the first to notice the laboratories' lack of self-acclaim. No banners celebrated, as they rightly could, that here is where scientists had developed the world's best-selling vaccines against measles, mumps, and rubella, or where researchers found and perfected the most popular prescription medicines against arthritis and high blood pressure.

When Scolnick drove off Sumneytown Pike for his first day of work, researchers in building 16 were already putting the finishing touches on a novel medicine to prevent heart attacks. Another team was testing a remarkably versatile molecule against parasitic infections that a few years later would cure the most common cause of blindness in Africa and also become the single most popular drug for farm animals and domestic pets around the world. Scientists were also developing new therapies against hepatitis, chickenpox, cystic fibrosis, osteoporosis, glaucoma, congestive heart failure, enlarged prostate glands, and even had underway a modest attempt to reverse male-pattern baldness.

Ed Scolnick had come to Merck's isolated West Point laboratories because he knew many of these things. For the previous ten years, the 42-year-old Scolnick had labored at the National Institutes of Health, rising from a precocious but unknown research scientist to join the ranks of the small but growing legion of internationally renowned molecular biologists. The difference in appearance between the NIH and Merck's West Point campus couldn't have been more startling. The thicket of high-rise NIH buildings where he had worked sat prominently on a hill off bustling, four-lane Wisconsin Avenue in the fashionable Washington, D.C., suburb of Bethesda, Maryland. It was well-known locally and around the world that inside the NIH compound some of the most important advances in medicine took root. Since 1972, in fact, Scolnick had directed a lab searching for the biological seeds of cancer. And in 1978, his lab pounced upon a gene he dubbed ras, whose pivotal role in human biology could only be guessed at the time. The name comes from "rat sarcoma," a kind of rodent tumor in which the gene was first found.

Four years later, Scolnick decided to leave the prestigious cocoon of the NIH because he had an ambitious plan. He wanted to turn his pioneering gene research into a revolutionary drug against cancer. Merck, he believed, was the place to do it.

In the years after Scolnick arrived at Merck, his research team and several dozen others in labs around the world learned that from its critical spot inside a cell, the protein produced by the ras gene acts as a central dispatcher, relaying chemical signals inside the cell that tell it when to divide and when to stop dividing. Scientists now know that if a cell's ras gene is somehow damaged, its cell-dividing signal can't be switched off. Ras (pronounced like "rat" but with an "s") soon became the paradigm cancer gene; when mutated it allows a normal cell to become an uncontrolled renegade, reproducing itself so many times that it eventually turns into a cancerous tumor.

Scientists say mutated ras genes may cause 20 to 30 percent of all cancers, including half of all colon cancers and a quarter of lung cancers. Many cases of breast cancer involve a mutated version of ras. So do many cases of lung cancer and brain cancer. And almost all cases of pancreatic cancer, a disease so pernicious it has defied all therapy, are the result of a damaged ras gene. As a result, ras was viewed by scientists like Scolnick as the perfect "all-purpose" cancer gene to attack. "It was, we felt, a kind of master switch for cancer," Scolnick says. "Turn it off and you turned off cancer."

Still, several years after Scolnick arrived at Merck, his enthusiastic hopes for quickly coming up with an innovative cancer drug faded. In a fateful meeting in 1989, Scolnick even suggested Merck end the project. "I thought we had run out of ideas," he says. "Luckily, I was persuaded [by others on the team] to try one more thing."

But while the project stumbled through the 1980s, Scolnick's fortunes at Merck did not. His scientific acumen and leadership qualities were quickly rewarded at the company. And as he steadily climbed through Merck's management ranks eventually to become director of the company's $1.5 billion-a-year worldwide research effort (and even briefly vie for the company's top job as chairman and chief executive), Scolnick quietly kept his small band of scientists on the trail of a treatment based on his ras cancer gene.

Indeed, it wasn't until thirteen years after Scolnick joined Merck that his cancer gene project, largely hidden from public view and nearly abandoned several times, was finally on the verge of producing a commercial drug. If the drug eventually succeeds—still a very large if—scientific historians may look back on its development at Merck as a major turning point in the war against cancer. For the first time ever scientists will have created a drug specifically designed to disarm a cancer-causing gene. Many cancer experts believe the experimental drug's emergence thus heralds the sort of targeted (or, in pharmaceutical terminology, rational) approach to fighting cancer expected to dominate twenty-first century medicine. It will be, scientists hope, the first drug to result from the hunt for cancer genes.

In June of 1993, Scolnick's ras gene laboratory at Merck, along with a competing team led by two esteemed Nobel Prize winners in Texas, reported a major milestone. Independently, the labs said they had found prototype drugs that, in test tube studies, appeared to do what Scolnick had dreamed: The drugs turned off the cell-dividing activity of mutant ras.[1] Fifteen months later, Merck showed that its experimental ras-blocking drug was able to shut down tumor growth in a small test in laboratory rodents.[2] And eight months after that, the same group electrified the cancer research community again when it showed that the prototype drug completely wiped out all signs of advanced breast cancer in mice specially bred to develop the disease. Moreover, the drug carried out its work without causing any of the harmful side effects associated with traditional cancer treatments such as chemotherapy and radiation.[3] The announcements alerted scientists that a major new strategy to combat cancer was being developed. And suddenly, Merck was plunged into a race against laboratories at a half-dozen major pharmaceutical companies that decided to chase after ras-blocking drugs of their own.

In a conversation a few months later, Scolnick acknowledged that the research findings, and Merck's apparent worldwide lead, had him excited but anxious. By now Scolnick was an important industrial executive whose public comments could spur or dampen the value of Merck's millions of publicly traded stock shares. "It's impossible at this time to predict" what effect the drugs will have in cancer patients, he said in a sober and cautious private conversation. "But I am thrilled by what we've accomplished to date. It's just that so much more needs to be done."

As director of Merck's drug discovery programs over the years, Scolnick had watched powerlessly as dozens of promising research leads for drugs against AIDS, heart disease, and other ills evaporated. As a result, he had learned from painful experience that the ras project faced a high probability of failure. The scientists, he said, might never make a version of the drug that was adequately tolerated by humans. The drug could be toxic in some unfathomable way. When finally put into people, the drug might not work as it did in petri dishes and in animals. "Right now, we are at an unbelievably exciting and frustrating stage of drug discovery," he said of the test compound. "We just have to be patient and wait."

But even if the ras-inhibiting drugs don't triumph over the many hurdles ahead, it was now apparent to Scolnick, to his competitors at other drug companies, and to some of the cancer gene pioneers that the dramatic shift in cancer research based on cancer genes had taken hold. "For the first time in my thirty years as a biomedical scientist I now believe that we will eventually conquer cancer," J. Michael Bishop, an esteemed biologist at the University of California, San Francisco, said in congressional testimony in early 1995, just as Merck stepped up its animal tests of its ras-blocking drug. "It may take several generations more of hard work, but the strategies for conquest are clear."

Indeed, the two decades of effort that led to the initial ras gene drugs suggest that developing new treatments based on the cancer gene discoveries is going to be a lengthy and risky enterprise. Even so, for biologists like Michael Bishop there no longer was any doubt that, as Vogelstein, King, Collins, Mark Skolnick and others pursued cancer genes, entirely new avenues of cancer treatments would be revealed. "The gene discoveries are at the core of every significant research effort [against cancer] these days," says John Minna, a leading cancer scientist at University of Texas Southwestern Medical Center. "The question still remains: Can we apply these [discoveries] to human disease? The ras drugs will be the first to test the concept."

Like many of his contemporaries, Ed Scolnick's route to biology research was not straightforward. Although he liked to tell friends that his name means "schoolboy" in Russian, Scolnick says he exhibited little that made him stand out as a young student in the middle-class Boston suburb of Dorchester.[4] Then a fifth-grade teacher persuaded his parents to push Scolnick ahead a grade and to transfer him to the prestigious Boston Latin School, where he excelled. In 1957, none of his friends or family were surprised when he was accepted by Harvard University.

"I didn't know what I wanted to do," Scolnick says. But then, after hearing a lecture on DNA, the structure of which had been discovered only a few years earlier, he was drawn to science and biology. "That's how you really understand what life's about," he decided. "That's what I'm going to do."

But Scolnick soon felt he was over his head. "I always wanted to do research," he says, "but whatever I tried to do in college and at Mass General [where he fulfilled his medical residency] was awful." It was a time of doubt and difficulty.

Then, fortuitously, he met Marshall Nirenberg, a veteran biochemist whose fame was legendary among the young scientists. In 1961, Nirenberg had helped decipher the meaning of the words spelled out by DNA's base pairs of nucleotides—the A's, T's, G's, and C's—that make up DNA. "Nirenberg had cracked the [DNA] code," says Scolnick. "We looked to him for guidance, and I was lucky enough that he took an interest in me."

From Nirenberg "I learned how to do research," Scolnick says. "He had a systematic disciplined approach to doing things. He preached the use of rigorous controls and record-keeping, something no one else had insisted on as being so important. Over time I realized that these simple but important tools helped me to become a good investigator."

In 1967, armed with an undergraduate degree in biochemistry and his M.D. and his residency completed, Scolnick joined a host of young men seeking refuge from the Vietnam War by enlisting in the U.S. Public Health Service as a novice scientist at the National Institutes of Health. Congress was stepping up its funding of cancer research, and the NIH's National Cancer Institute was soon awash in new money for new labs, equipment and young, enthusiastic researchers willing to tackle the mysteries of the disease. For them, the NCI was a safe and nurturing haven where they could serve out the war dressed in white lab coats, not green combat fatigues.

In 1969 Scolnick joined the NCI lab of George Todaro, an intense and soft-spoken man a few years older than Scolnick. Todaro was trying to use new genetic tools to determine whether viruses cause cancer. For the next few years, Scolnick and the Todaro group picked apart as best they could a virus known to cause sarcomas, or tumors of connective tissue, in rats. Then in 1975, Scolnick was put in charge of the cancer institute's new tumor virus section, with a long hallway of biology labs at his disposal. He was to find the component(s) of the virus—the genes—that causes cancer.

At about the same time came riveting news from across the continent. Two University of California researchers, Harold Varmus and Michael Bishop (whose testimony in Congress twenty years later was cited earlier), found that a particular chicken cancer was triggered by a mutated gene that somehow had become incorporated into the virus's own genetic machinery.

Moreover, the California scientists flabbergasted their colleagues when they made still another pronouncement. They found the same cancer-sparking gene, or ones almost identical with it, in cells in every species they looked at, from fish to humans. "When that happened we were certain we'd found an important gene simply because so many species of animals had retained it over the millions of years of evolution," Bishop says. "It had an important role, but what it was we didn't know."

Their research set off a sprint among biologists for other so-called "oncogenes," or cancer-causing genes. (Bishop and Varmus were honored with the Nobel Prize in 1989 for their trailblazing work.[5])

The scientists concluded that at some point in the past, the avian virus had somehow plucked the extra gene from within chicken cells. Then, tucked inside the virus's interior, the extra gene incited tumor growth when the germ infected other chickens. Exactly how this gene-studded virus caused cancer was still a mystery. "I knew nothing about avian viruses; we were studying cancer-causing viruses in rats and mice," Scolnick says. "We wanted to see if the same rules applied."

In 1978, Scolnick's labs identified the gene and named it ras. Within a few years, scientists using similar strategies isolated several other oncogenes. (By the mid-1990s, several dozen oncogenes had been found.) All had a similar profile. Their structure was conserved across an unusually wide range of species, and the proteins they produced appeared important to cell growth. Most especially, the researchers showed that simply mixing the genes derived from the animal viruses into a batch of cells in a test tube was enough to generate tumor growth.

But how did one gene wreak so much havoc? For the next few years, several research teams set about finding out.[6] "We were pretty convinced that, for the first time, we were looking for the molecular basis of cancer," says Robert Weinberg, a leading cancer gene researcher at the Whitehead Institute for Biomedical Research in Cambridge, Massachusetts.

In 1982, Weinberg's lab produced a result so unbelievable even it had trouble accepting it at first. Within the gene's thousands of base-pair nucleotides, just one letter out of place caused the normally benign ras to become an oncogene. Weinberg called the replacement a "single point mutation." It was a DNA alteration so slight it took almost four years to ferret out. But it was so powerful that by itself this one change rearranged the molecular course of a cell's life, transforming it from a benign citizen into a dangerous one.

The short, balding Weinberg, whose descriptive speech earned him nearly as much acclaim as his science, was universally hailed. In a commentary accompanying publication of the point mutation, the editors of Nature magazine praised it as "one of the most startling discoveries in the long and frustrating search for an understanding of cancer."

Few scientists doubted the importance of ras or Weinberg's discovery. For his part, Ed Scolnick already had concluded that he had gone as far as he could in a government-funded laboratory system. Tall and broad-shouldered, his chest bent slightly forward as he walked, the dark-haired, basement-voiced Scolnick was not easily ignored when he strode into a room, sternly argued a hypothesis, or gave no-nonsense directions to his staff. He knew he could join the oncogene-hunting fray. But he'd already found his gene. He wanted to do something bigger and different. Scolnick by now realized he hadn't become an accomplished scientist by treading down well-worn paths; it was time to move on.

"Even by 1980, I began to think about what I wanted to do for the next ten or fifteen years," Scolnick says. "I came to realize that I wanted to try to develop the [ras] discovery and others like it into new types of [cancer] therapy and diagnostics. Nobody else was thinking that way. I looked around and realized the labs at NIH, or at any university post no matter how well-endowed, were not going to provide the resources needed for the job ahead. I was looking for a way or place to do this kind of work. Then Merck called."

More precisely, one day in early 1982, Scolnick received a phone call from P. Roy Vagelos, a man he had barely met but knew from reputation. By 1982, Vagelos was on his way to becoming the latest, and quite possibly the most successful, heir to a line of innovators Merck traces back to 1668 when Friedrich Jacob Merck bought an apothecary called "At the Sign of the Angel" in Darmstadt, Germany. In 1816, the Merck family opened a factory, and over the next few decades it was one of Europe's principal producers of morphine, codeine, and cocaine for medicinal use. In 1891, 23-year-old George Merck was sent by the family to run a newly started branch of the thriving business in the United States, eventually opening a research and manufacturing facility in Rahway, New Jersey, then a rural town about 10 miles west of the bustling New York harbor.

Under Merck's son, George W., the company fashioned a reputation for conducting the kind of research generally found only in universities. Merck aggressively courted professors by promising to let them publish findings in esteemed research publications at a time when most companies jealously guarded their "secret" inventions. In the 1930s, George W. built a quadrangle of red-brick colonial-styled buildings at Rahway, complete with ivy, to help recruited academics feel at ease. It was at the Rahway campus that a young chemist, Lewis Sarrett, spearheaded a decade-long project that in 1944 culminated in the synthesis of cortisone. One of the first of the modern-day "wonder drugs," cortisone and its related compounds soon transformed the treatment of seventy-five conditions such as arthritis and allergies, and propelled the search for other anti-inflammatory medicines. The cortisone research popularized the name "Merck" among doctors and patients, and it soon became synonymous in the public mind with important, disease-fighting breakthroughs.

During this time Merck's drug discovery labs were led by Max Tischler, lionized to this day inside Merck as setting the company's high standard for basic research. With him at the research helm, the company and its labs burnished its reputation, producing the first penicillin used by an American patient, and uncovering B_{12}, the vitamin to combat anemia. After the war, Merck collaborated with Rockefeller University researcher Selman A. Waksman, helping him isolate the antibiotic streptomycin, for which he was awarded the Nobel Prize. So great was Merck's fame that in 1952, *Time* magazine put George W. on its cover, accompanied by the headline, "Medicine is for people, not for profits." In 1970, when Tischler retired as head of laboratories at Rahway and those acquired in 1953 in West Point, Pennsylvania, his name was on the patents of ten of the top-selling drugs of all time.[7]

Still, when Roy Vagelos first visited Merck's labs in 1975 he was thunderstruck by the "primitive" approach that formed the foundation for the company's drug research. Mighty Merck hadn't produced a significant new drug in several years; its labs were in the midst of a terrible research drought. Vagelos believed he knew why. Confident and competitive, Vagelos vowed to himself that if he took the job as head of research, the labs would have to agree to make significant changes. "I couldn't believe the way Merck, of all companies, was going about things," he later recalled.

As a teenager, Vagelos had waited tables at the Irving Street luncheonette owned by his Greek-immigrant parents just six blocks away from Merck's Rahway operations. Enthralled by the scientific talk he overheard from Merck employees who ate lunch at the restaurant—and by the burgeoning fame emanating from inside the research gates—Roy set his sights on medical school, first working his way through the University of Pennsylvania and then Columbia University School of Medicine, where he earned an M.D. He wasn't interested in going into private practice. "In research you have the possibility of helping so many more patients than just your own," he often explained. After ten years at NIH he went on to Washington University School of Medicine in St. Louis, where his research into the way the body makes use of fatty proteins called lipids gained him membership into the select National Academy of Sciences.

Encouraged by a former mentor to interview for the top research post at Merck, Vagelos was dismayed when he toured the company's labs. Drug research was largely still a trial and error process. Chemists would churn out hundreds and thousands of compounds, which researchers tested in animals or in test tubes to see if any had a desirable effect. Drug discovery, like all scientific inquiry, is a combination of hard work, insight, technical adroitness and a large dollop of good fortune. Merck, Vagelos throught to himself, was relying too much on luck.

Vagelos believed it was possible to exploit the advances in biology that were exposing the step-by-step biochemical interactions of proteins and enzymes that underlie human disease. The trick, he argued, was to identify biochemical components in a disease pathway, pick one that appeared vulnerable to outside attack and then search for chemical compounds, drugs, to deactivate it.

Vagelos was convinced that enzymes, powerful catalysts of chemical reactions inside the body, were the key. Disable an enzyme that propels the chain of biochemical events of a disease, he preached to his new research troops, and you have a potentially powerful new drug. Within months of joining Merck, Vagelos ordered research teams to abandon any project that wasn't designed to identify and interdict the enzymatic machinery driving a disorder, from hypertension and arthritis to diabetes, cancer, allergies and even the common cold.

Over the twenty years that Vagelos was head of research, and then chairman and chief executive of Merck, the enzyme-inhibiting tactic produced a basketful of prescription drugs that pumped billions of dollars into Merck. These medicines included Vasotec, the world's best-selling drug for high blood pressure; Mevacor and Zocor, innovative prescription medicines to lower cholesterol levels in the bloodstream and prevent heart attacks; Primaxin, an antibiotic that helped transform the treatment of infections involved in cystic fibrosis; and Fosamax, a new type of drug for preventing osteoporosis, the bone thinning that can cause all sorts of health problems for aging women.

Of course, some projects were busts. But by the middle to late 1980s "every major drug maker had embraced the idea of finding a target, such as an enzyme, in the disease process and blocking it," said Charles Sanders, a contemporary of Vagelos, who led research at Squibb and at Glaxo, two major Merck competitors. "We all learned from Merck's success."

Among Vagelos's many attributes was a special knack for attracting and motivating top-rank scientists, many of whom previously had been unwilling to leave academia. "Roy could be hard and demanding and you knew he expected the very best," says Manuel Navia, who worked in Merck's West Point labs in the 1980s. "But he also knew when to back off, when to praise. His ideas were exceptional. It gave him a lot of credibility. He was a very involved, very exceptional leader."

Under Vagelos, Merck was once again king of the pharmaceutical industry hill. Beginning in 1987, and then for six consecutive years, Fortune magazine named Merck the nation's "most admired company." Indiana University chemist David Paisely, then a securities analyst at Merrill Lynch, joked that "Merck's name is so distinguished the company could bottle crap and doctors would write prescriptions for it with both hands."

By the early part of the 1980s, Vagelos already had decided in his characteristically sanguine manner that Merck's labs should use the new "rational" (enzyme-targeting) approach to tackle every disease under the sun. Except for a few crude treatments, however, Merck didn't market cancer medicines, still doesn't today, and it had no cancer studies underway. The small batch of cancer projects it had were halted in 1967 when executives decided too little was known about the disease and too little money could be made in the field. But by the early 1980s, Vagelos was well aware of the fuss over oncogenes, and believed, like others, that oncogenes might help scientists map the cancer pathway, ultimately identifying the kind of enzyme targets Merck could inactivate. Back then, most cancer drugs under development at other companies and research centers worked by delivering lethal poisons to a cancer cell, often causing side effects as dangerous as the disease. "It was an old-fashioned, shotgun approach to making medicines," Vagelos says. "It wasn't my way of doing things. I thought the oncogene approach was more directed, more logical. It was trying, finally, to make sense of the disease for which almost nothing as regards its basic cause was known."

Vagelos realized that Merck and other drug makers were woefully behind in exploiting new gene-based technologies. His labs were ill-equipped to explore the oncogene pathway. Vagelos had met Scolnick briefly at the NIH and had taken an interest in his career, including his name among those he might someday lure to Merck. When Vagelos heard in 1982 that Scolnick had spoken to officials at Massachusetts General Hospital in Boston about an opening there, "I knew Ed was movable," Vagelos says. "Ed was a very hot item," as Vagelos puts it. "We were lucky to get him."

Scolnick quickly realized that most of his time as director of Merck's main research facility in West Point, Pa., would be taken up learning Merck's drug discovery system and building up its basic biology research. He handed off principal responsibility for exploring ras to Jackson (Jay) Gibbs, a 27-year-old with a newly minted

Ph.D. in biochemistry. Gibbs was assigned to work with Irving Sigal, a young molecular biologist newly arrived from Harvard Medical School. "Irving and I worked together very closely," says Gibbs. "But at the end of every week, I'd write Ed a memo or send him a copy of my lab notes to read at home over the weekend. On Monday, I'd find a detailed critique waiting for me when I got to the labs.

"Back then, we naively thought maybe we'd found the one or two genes that when mutated caused all cancer," says Gibbs. "We knew we'd have to be lucky to find a drug quickly. But we were very excited because we realized at Merck we had a real chance of blocking a key cause of cancer. Now we know there are probably a hundred oncogenes, and that, of course, the process is much more complex."

Between 1982 and 1985, the small group of less than a half-dozen researchers toiling away in the West Point labs tried to chemically isolate the overactive protein produced by mutated ras. They were convinced the project fit the new Vagelos model of identifying a target in the disease pathway and then selectively stifling it. "Our sense was that a ras-blocking drug would be potent, but, as important, it would be safe," says Gibbs, a tall sandy-haired man who selects his words carefully and delivers them, even when excited, slowly and with little emotion.

Scientists had learned that for ras to trigger cell division, the protein first was biochemically "turned on" when a tiny molecule called GTP fastened to it. When it was bound in the same spot to another similar molecule, called GDP, ras remained "off." Researchers realized that under normal circumstances these molecules were constantly switching ras back and forth through "on" and "off" cycles. But, they found, mutant ras appeared to be bound only to GTP. Unable to recognize the "off-cycle" molecule, it remained in constant embrace with the "on-cycle" molecule.

To shut off ras, the Merck researchers devised a standard drug searching experiment in a test tube. At random, the scientists tossed into the test tube chemicals they pulled from the company's extensive library of compounds. They hoped one of their chemicals would nudge aside the GTP "on" molecule and mimic the muting action of GDP. During the project's first three years, "we must have tested 100,000 inhibitors," Gibbs says. "Nothing worked."

In 1985, Roy Vagelos was rewarded for revitalizing the company's research. Merck's barren pipeline was now stuffed with several potential blockbusters, and rising sales and earnings placed Merck first among U.S. pharmaceutical makers in prescription drug sales, up from second place where it had slipped in the mid-1970s. In April, just before the company's annual meeting, Merck's board of directors handed over the company reins to Vagelos, naming him president and chief executive officer, the only doctor and scientist to lead a U.S. drug maker. Vagelos then chose Ed Scolnick to succeed him as head of drug discovery.

By now, the labs of Sigal and Gibbs were among a dozen or so research programs regularly leapfrogging one another with new insights into the ras pathway. Several teams found evidence of Bob Weinberg's point mutation in cells removed from breast, lung, colon, ovarian, and pancreatic tumors, suggesting the damaged version of the protein was even more important than previously suspected. "As the data rolled in it became apparent that ras was the most commonly involved of the oncogenes, and that it was the target to go after," Scolnick says.

Then, in the summer of 1987 a scientist from a small California biotechnology company dazzled a research conference outside Washington, D.C.; he'd found a way to cripple ras. The researcher, Frank McCormick, told the gathering that his lab had found evidence that ras was activated by joining with another unidentified protein he called GAP. "It was a stunning discovery," says Gibbs, who heard it for the first time while sitting in the meeting's audience. "McCormick had evidence of the first protein to interact with ras. It crystallized for eveyone what we needed to do." Stop the two proteins from binding and, the scientists believed, you might stop ras.

But first the scientists had to isolate GAP's exact structure, and for months following McCormick's disclosure every ras lab raced to find it. By the fall of 1987, Gibbs had purified a bovine version of the protein and by the following spring he had cloned the animal gene. Several weeks later, McCormick cloned the human counterpart. "Gibbs's lab was more efficient at chasing down the gene," says McCormick. "I've got to hand it to Jay. He showed he's a great biochemist."

With GAP now in hand, the race entered a new lap as Merck's team and others, including one led by McCormick, tested compound after compound, hoping to find one that would inhibit ras by fitting snugly between the points where GAP and ras locked together. Then in late December 1988, just four days before Christmas, disaster struck in the form of a terrible human tragedy. Pan Am's Flight 103 from London to New York was torn apart by a terrorist bomb over Lockerbie, Scotland, killing all on board. Gibbs heard about the explosion on his radio while driving home from the labs. Later that evening, Scolnick telephoned. "Irving was scheduled to be on that flight," Scolnick said, referring to Gibbs's boss, Irving Sigal.

"The next day it was confirmed," Gibbs says. "We were devastated. But to Irving's credit he had put together a strong group. We were a pretty stunned bunch of people but we kept moving on. Maybe it was our way of dealing with his death."

But no amount of dedication to Sigal's memory could solve the scientific problem they faced. Into the summer of 1989, the researchers tested tens of thousands of potential ras-GAP inhibitors without success. "In retrospect, based on what we now know, we never should have wasted the time," says Allen Oliff, an oncologist and biochemist from Memorial Sloan-Kettering who now commanded the ras project. The reason for failure, the Merck team soon learned, is that proteins fasten by linking across a broad bit of molecular geography. "The protein-to-protein binding domain is just too large," says Oliff. "It would have required a chemical much too big to get inside a cell, where ras and GAP were joined. We know now that you can't block most protein-to-protein binding reactions inside a cell."

Frustrated by the science and depressed by Sigal's death, a gloomy Scolnick believed the ras research had run its course. In the autumn of 1989, he called together the cancer group and in a small conference room outside his office in West Point he decided to tell the researchers to move on to other targets. "By then Merck had given seven years to ras," Oliff says. "We had other cancer genes that looked exciting. We talked about abandoning the [ras] project."

That is, until Gibbs asked the assembled group if his lab could try "one last thing." Over the previous few months, Gibbs had collected several research reports suggesting that still another step in the ras pathway may have been uncovered. Until the meeting with Scolnick, Gibbs hadn't shared the unfolding information with anyone. "It was an early but exciting idea," says Gibbs. By 1989, Merck was completely immersed in Roy Vagelos's enzyme-blocking strategy for finding drugs. "Our chemists had become very adept at designing enzyme inhibitors," Gibbs says. Thus, Gibbs told those assembled in Scolnick's cramped conference room, the "last thing" he wanted to try before shutting down the ras project was to hunt for a lipid-fusing enzyme—although he couldn't be certain that it even existed—and then set loose the company chemists to find a compound to block it. Deprived of the nourishing lipid, or fat, by a drug inhibitor, Gibbs figured, the starved ras would be too feeble to carry out its grim task.

Merck shut down every other ras-related cancer project and put all its effort into pursuing the enzyme and an inhibitor.

But unknown to the Merck team, another set of researchers in Texas were pursuing the same approach. The other research group couldn't have been a more daunting rival. Based at the University of Texas Southwestern Medical Center in Dallas, the lab was run by Michael S. Brown and Joseph L. Goldstein, two men who had won a Nobel Prize in biochemistry in 1985 for picking apart the pathway by which cells process cholesterol, which, of course, is a type of lipid ferried through the bloodstream.

Throughout the rest of 1989 and into the spring of the next year, Gibbs and his laboratory tried to sequester the putative enzyme from the soup of substances that surround ras inside a cell. But, they soon were surprised to learn, they weren't able to match the Dallas lab's expertise with lipid enzymes. In February, the No-bel laureates revealed that they had jumped into the ras race when they published a research article in *Nature*[8] describing exactly how the unknown enzyme was critical to producing cholesterol and, also, to activating ras. In the article, Brown and Goldstein dropped a rhetorical bomb: "Inhibitors [of the enzyme] could be useful on a chronic basis to suppress the growth of certain tu-mors."[9]

Four months later, the Dallas group declared victory, isolating and purifying the enzyme, which they named farnesyl tranferase, or FT. They were able to explain how FT worked, including dis-closing a crucial discovery that meant that it was possible to in-hibit the coupling of ras and the enzyme with a similarly small chemical, a drug.

"Inasmuch as oncogenic ras proteins have been identified in a variety of human tumors, the possibility arises that the inhibition of this [enzyme] will be of therapeutic benefit in the treatment of such diseases," the Dallas team wrote.[10]

Soon afterward, Scolnick telephoned the Dallas researchers, suggesting they reprise a previous collaboration with Merck. "Ap-parently [Brown and Goldstein] had other plans," Oliff says. "Later we learned they already had hooked up with Genentech [a biotechnology company based in San Francisco, California]."

Within days, Scolnick designated the ras project a full-blown drug search despite misgivings about the potential toxicity of a drug that blocked the FT enzyme. At a scientific symposium, sev-eral researchers argued that because the enzyme was ubiquitous in human cells a drug that deactivated it might cause mayhem else-where in the body. "I decided not to go after an FT inhibitor," says Frank McCormick, who had formed a new biotech company out-side San Francisco to continue his search to subdue mutant ras and other oncogenes. "I was convicned that an [enzyme] blocker would be toxic. It didn't make intellectual sense."

McCormick worried about something else, too. He knew that any drug that blocked the enzyme wouldn't actually kill the tumor, as did poisonous chemotherapy medicines or radiation. Instead, as long as the drug was administered, it would only hold cancer growth in check. In other words, the drug would have to be constantly given to patients to suppress their cancers. No drug like that had ever been developed before against cancer.

But Scolnick swiftly dismissed concerns brought to him by Oliff and others. "It just smells right," Scolnick said of the enzyme-blocking strategy. He ordered all correspondence regarding ras between Merck and Dallas terminated. Even so, Oliff and Gibbs were unable to muster a full-fledged attack against ras. At giant Merck, where dozens of research groups were targetting enzymes involved in a host of disorders, competition "for adequate resources was a constant problem," says Oliff. "For the longest time we simply couldn't get our hands on enough of everything."

For the next two years, independent of one another, a handful of chemists drawn to the project at West Point and an even smaller number toiling at Genentech designed dozens of tiny molecules, searching for one that could latch tightly onto the four amino acids where ras and the enzyme normally joined. Finally, in late December of 1992, Merck's chemists won the race, producing a candidate drug, designated L-731,734, that sat snugly in the coupling site, blocking the enzyme from attaching the lipid to ras, and preventing ras from anchoring to a cell's perimeter. In a remarkable experiment, the Merck team was able to tranquilize a batch of fast-growing cells fueled by mutant ras by injecting the prototype drug into their midst.

Working feverishly to beat the Dallas team into the scientific literature, Gibbs and Oliff scurried to get news of their drug discovery into print, mailing off a brief report to *Science* magazine at the end of March 1993. Within days, Brown and Goldstein, and their colleagues at Genentech heard rumors of Merck's find. As it happens, just a few months earlier the Dallas-Genentech collaboration also had produced a powerful enzyme blocker. Three weeks after *Science* received Merck's report, the editors received another research article from the competing Dallas-Genentech team. It claimed the discovery of a totally different molecule that worked as well as Merck's.

The two reports were published simultaneously in late June, and were greeted by a raft of publicity, including an article in *The Wall Street Journal*.[11] In it, James C. Marsters, Jr., the chemist who designed the Genentech drug, said: "This enzyme is an Achilles heel of the ras protein, and our thought was that if we could block it, we could inhibit the action of the ras oncogene on cells." The article was the first time Merck or Genentech publicly acknowledged their pursuit of a cancer therapy based on disarming the ras oncogene.

But inside Merck, excitement over the drug was muted. "We knew we still had to prove that it worked beyond the test tube," says Oliff. "By the time it was published in *Science* we already were gearing up to try it out in [laboratory] rodents."

The researchers set up an experiment to determine if the prototype drug could reverse the growth of a tumor inside a mouse. The scientists squirted into the skin of the mice a syringe-full of human cells containing mutant ras and other mutated cancer genes. Within a few days, these cells began to form tumors. A few days after that, the Merck scientists began injecting one group of the mice with the prototype drug, leaving another group untreated. About a week later, the researchers surgically removed the tumors and compared their size. The tumors taken from the treated mice were a half to one-third the size of the tumors from mice not given the drug. Moreover, the tumors shrank to about half the size of those tumors treated with a commonly used cancer chemotherapy.

While the anti-tumor activity in the mice was important, the scientists noticed something even more profound and mysterious: The prototype drug produced no discernible side effects. "We were, quite honestly, a bit flabbergasted," says Oliff. "We hoped and assumed the drug would have an effect on the tumors and we were, of course, very gratified when the results became apparent. But, in regard to toxicity, we didn't know what to expect. Apparently, we think, tumors make use of a lot more of the enzyme than do normal cells. At the dose we were using, normal cells just don't seem affected by enzyme inhibition."

Scolnick felt relieved and vindicated. "Every time I asked Ed how it was going, he played it down," says Vagelos. "It wasn't until the mouse study that he finally got excited or, at least, that he finally let me know we had something important. He had so much invested in making it happen, I don't think he wanted to get his hopes too high."

But quietly with his staff Scolnick was ecstatic. Presented the mouse study findings one day in late spring of 1994 at West Point in a briefing by Oliff and Gibbs, Scolnick leaned back in his chair and allowed himself a long moment of pleasure. "From the beginning," he told Gibbs later, "there's been something magical about this research."

Still, Merck had a great deal more to do. By late 1994, about a half-dozen major drug companies were rushing to create their own enzyme blockers. At Merck, more than two dozen chemists were assigned the difficult task of refining the prototype drug into one that could be tolerated by people. It wasn't until early 1997, however, that Merck's chemists finally fashioned a set of drugs considered suitable for testing in cancer patients. Even then, the experimental compounds faced months and months of animal studies before they could be tried in people.

Indeed, by then, researchers at several large drug companies and small biotech outfits were chasing after drugs to block still other proteins identified as important messengers in the ras signaling pathway. "The ras pathway is becoming an unbelievably fertile area for researching new cancer drugs," says Joseph Schlessinger, a New York University biochemist whose small company, Sugen, Inc., was targeting several proteins that affect ras.

And many other questions lay ahead. For instance, scientists still weren't certain how a drug that merely freezes a tumor in place would be administered. Some researchers argued that once held in check the tumors might become much more vulnerable to traditional chemotherapy. Perhaps, they said, given the drugs' apparent lack of side effects, patients would be willing to take the drugs for many years.

"If we can get the drugs into patients, I think the data will show the concept is right," Scolnick said at the time. "Right now, we're at an unbelievably exciting and frustrating stage of drug discovery. We just have to be patient and wait."

But even if the ras-inhibiting drugs don't triumph over the many obstacles ahead, it was now apparent to Scolnick and to some of the cancer gene pioneers that a dramatic shift in research based on identifying cancer genes had taken firm hold.

15: First, Do No Harm

Several days after the breast cancer gene discovery was announced in September of 1994, 35-year-old Rachel Stein set off on a personal journey into the brave new world of genetic testing. Stein, a suburban New Jersey health care professional and mother of two toddlers, was unprepared for the initial news reports. The press accounts hailed the isolation of BRCA1 as an important first step in someday curing the cancer. But each media story also included an additional, unexpected and, to Stein, a truly hopeful detail: scientists could now produce a test to detect who in the population carried a relatively common cancer-causing gene, one that was responsible for much of the 5 to 10 percent of breast cancers that are caused, in large part, by inherited factors.

Stein remembers feeling a quiet celebration at hearing the news. Finally, after years of worry and uncertainty that had shadowed her young life, there now was a way to find the truth about her family's harrowing past and, quite remarkably, whether it would be repeated in her life, and the life of her young daughter, too. "I'd been living with so much dread for so many years," she says. "I wanted to know. I guess I felt like I really needed to know. Did I have the gene?"[1]

After a terrible series of family tragedies, Stein had become convinced that she carried a deadly gene, and that it was only a matter of time before she contracted cancer. When Stein was just 22 years old and graduating college, her 45-year-old mother was diagnosed with breast cancer. Four years later, her mother succumbed to the illness.

Stein's mother was the third of three sisters to perish from breast cancer. Moreover, Stein's maternal grandmother had died of cancer, too, at the young age of 40. All the women in her mother's family passed away before they reached age 50.

"For years my cousin [the daughter of one of the two deceased aunts] and I talked about our risk, and what we ought to do about it," Stein says. "As we got older, Staci [the cousin] and I talked about protecting ourselves. We talked about removing our breasts." When the breast cancer gene's discovery was reported, and there was now the possibility of a gene test, "I knew right away that's what I wanted," Stein says. "It held out the possibility of giving me and my daughter a way out."

Like Stein, thousands of other women had long wondered if the cancer stalking the female members of their families was simply a terrible fluke, or due to some dark but unknown force, long hidden, but now, with the dramatic cloning of BRCA1, finally exposed. But whether doctors would be able to use a test to predict a person's future health was a subject immediately caught up in controversy.

The arrival of the new gene test on the American medical scene was noted by many social commentators as the dramatic commencement of an entirely new era in medical diagnostics, one certainly full of great promise, yet also fraught with unexpected peril. Medicine had leaped into the twenty-first century world of technology, commentators asserted, when doctors will have at their disposal a battery of similar gene tests to identify an individual's inborn susceptibility to a wide range of common conditions such as asthma, heart disease, many cancers, diabetes, arthritis, obesity, allergies, depression, and even baldness.

With each test would come the possibility of preventing, avoiding or limiting a malady's impact, perhaps through life-style changes or through some as yet undiscovered medical intervention. But these tests will bring many of the same disturbing ethical dilemmas associated with the breast cancer gene test.

From the time of the breast cancer gene's discovery in late 1994 until late 1996, as Stein struggled to get the answer she wanted, she found herself an unintentional participant in a raging battle. Some geneticists urged cancer specialists to move cautiously in making the test widely available, arguing that much research was needed before the test could be responsibly released to a public and to a medical community unprepared to deal with the sophisticated subtleties of genetic risk. At the same time, breast cancer family members clamoring for the test were enraged by efforts to slow the test's release.

By late 1996, the gene test was embroiled in a debate over its sale to the public and to doctors by profit-making ventures. Two private companies, Mark Skolnick's Myriad Genetics in Utah and another gene-testing firm, called OncorMed, in Maryland, began marketing the test to cancer doctors and medical centers, sending the providers letters and brochures describing the new test's benefits for women believed at risk of the disease. Another for-profit medical concern even began brashly advertising the test in newspapers and magazines, triggering howls of concern. Several national breast-cancer advocacy groups publicly attacked the companies' tactics, alarmed as they were that a promiscuous and unregulated use of the test would end up frightening women with a social and medical stigma that threatened their employment, their ability to get or retain health insurance, or even their emotional health.

"We are entering a real danger zone with this test," said Michael Kabak, a geneticist at the University of California, San Diego, who regularly participates in panels assessing the risk and benefits of genetic testing. "Just because we have mastered the extraordinary science making it possible to identify who carries a mutated gene and who does not doesn't mean we have the right to be testing right now for its presence, or that we are so smart that we know how to handle the information it provides."

Francis Collins, his role as director of the National Institute for

Human Genome Research[2] by this time giving him de facto responsibility for national gene policy, said he believed information produced by the gene test was "potentially toxic" to women and their relatives, and he publicly called for federal legislation and oversight to regulate its sale and use. "One of my fears is that we will commit so many egregious errors early on that the American public will decide that they do not want to have anything to do with this technology," Collins told a special forum on gene testing convened in San Francisco in November 1996. "We don't need a genetic thalidomide."[3]

But Skolnick and others called such worries paternalistic. "This test will save lives, possibly hundreds of thousands of lives," Skolnick said to a gathering of doctors and health professionals in New York City a few months later. "The alarmists who would keep this test away from women are doing more harm than good."

Kenneth Offitt, an oncologist at Memorial Sloan-Kettering Cancer Center in New York, says that in his experience, "This test can be empowering for a woman, especially for someone who has lived in a family where for so long nobody knew how or where the cancer had descended. This helps explain the unknown, and helps people make choices about their future they wouldn't otherwise be able to make. But because it is such a powerful test, yes, it must be handled very carefully."

Central to the rising concern are several hard facts of science. Anyone in a cancer family who does not have the mutant version of the family cancer gene is indeed free of the risk of getting the gene-caused disease, a risk that scientists calculate to be about 85 percent over a lifetime for someone carrying a gene mutation. This 85 percent risk figure is based on analysis of disease frequency among gene carriers in large families studied as part of the original gene hunt. In other words, 85 percent of the women in these research families who harbored a mutation eventually developed breast cancer. And about 40 to 50 percent of women with the mutation developed ovarian cancer, meaning that the lifetime risk of ovarian cancer for a breast cancer gene carrier was about 40 percent. Nonetheless, a noncarrier of a BRCA1 or BRCA2 mutation still retains the same 8 to 12 percent risk that all women have of getting breast cancer sometime in life. That is, about 8 to 12 percent of all women, no matter their family history, eventually develop breast cancer. Someone without the cancer-causing gene is not free of all danger of developing the disease.

"We've run into cases where women who test negative for a mutation have been led to believe by their doctors or whoever conducted the test they don't need to be as vigilant as other women, that they needn't continue doing self-exams and undergoing mammographies as is recommended for all women," said Jill Stopfer, a counselor specializing in cancer genetics at the University of Pennsylvania Medical Center, Philadelphia. "In other words, whoever was providing the test either misled the women or was unable to clearly explain what a negative result means."

In April 1997, a surprising research study from the National Cancer Institute confused matters even more. Scientists there studying a large sample of women in the Washington, D.C., area reported that carriers of BRCA1 might only have a 56% chance of getting breast cancer in their lifetime and a 16% chance of developing ovarian cancer, percentages far lower than the estimates being used by Myriad and other promoters of the gene test. If the NCI research was correct, it meant that gene carriers had about a 50-50 chance of developing breast cancer. Mary-Claire King and Mark Skolnick both attacked the study as flawed, but Francis Collins and others at NIH disagreed, saying the new lower risk estimate might be closer to the truth.

The lifetime risk of getting cancer is a critical question, and King and others began assembling their own population studies to resolve the debate. Knowing the precise risk is important, because, by 1997, the options available to women found to carry the gene were limited to unacceptable, primitive options. The only acceptable medical intervention offered to gene carriers is explosively controversial: It is the surgical excision of all breast tissue—a so-called bilateral mastectomy—as well as the removal of the ovaries—an oophorectomy. As a storm swirled about the new data and the treatment options, there was only preliminary evidence that prophylactic surgeries prevented cancer, especially if the surgeon inadvertently left some breast tissue behind.

And, of course, while scientists hope and expect that having the BRCA1 gene in hand will lead them one day to new and powerful drug and other therapies to block the gene's cancer-causing effects (as illustrated by Merck's lengthy fight to thwart the ras gene), no new drugs targeted specifically against BRCA1 are expected in the very near future. BRCA1, in fact, was turning out to be a particularly difficult gene. Three years after its discovery scientists only had the first inklings of how the normal gene worked, or how a mutated version inspired cancer.

"So, we are left in a very tough spot for God knows how long," said Michael Kabak, the San Diego geneticist. "We can tell a woman who tests positive that she is likely to get cancer, but we can't tell her if she will get it for sure, or when it may occur. We can't tell her if she will get breast cancer or ovarian cancer or both. We can't tell her if her cancer will be prone to spread or slow to develop. In other words, we can only offer her the option of removing her organs, and we can't even tell her that that is a foolproof therapy. It's a terrible spot, to be told your future risk, that you have a high probability of developing disease, and not be given any other medical advice that many would consider very acceptable."

An editorial writer in the influential *New England Journal of Medicine* wrote that "Genetic testing [for BRCA1] will be clearly beneficial only when we have effective interventions and can guarantee protection against discrimination and social stigmatization."[4]

For many the discovery of the breast cancer gene produced a short-lived euphoria, indeed. Back in Michigan, Susan Vance and her sister Janet heard news of the isolation of BRCA1 and wept. "I cried and cried," says Susan. "I'd hoped and prayed so long for this. I believed a cure was now possible, maybe very soon." Adds Janet: "I thought about my daughter and nieces and cousins, that maybe this meant there would soon be some better option for them than removing their breasts and ovaries. Maybe they'd have opportunities I didn't have."

But within weeks, "we learned that if the gene was going to help produce better treatments, or even a cure, it was going to take a long, long time," says Janet, clearly disappointed by the hard realities of scientific progress. "We were overjoyed by the discovery, but my daughter and I were still left with hard choices. Nothing about finding the gene changed that."

Yet, for Rachel Stein and her 34-year-old cousin, Staci Mishkin, the BRCA1 discovery held out a bright glimmer of hope after years of horror at the premature death of loved ones. "My [maternal] grandmother died when my mother was only 12 years old, and it affected my mother's outlook on life," says Mishkin. "She grew up without her mother, with anxiety and insecurity and fearfulness. And then when I was in kindergarten my mother's oldest sister died. Both those deaths had a great impact on me, too. My mother was first diagnosed when I was 14 and she died when I was 25. These deaths shaped my life. I didn't want that repeated for my kids if I could help it."

Rachel Stein contacted a medical research center near her home known for studying familial cancers. But the scientists at the medical center said that to examine her gene they first needed to know precisely where along the lengthy stretches of DNA in BRCA1 or BRCA2 a specific mutation existed. This, they said, could only be learned by examining cells extracted from the blood of an affected woman in Stein's family, from either her mother or from one of her aunts or grandmother. But because all the women had died, the scientists said they couldn't perform the test at that time.

"I was very frustrated," says Stein. "I was totally prepared to hear the news either way. I wasn't prepared for the delay."

In Salt Lake City, Mark Skolnick and his colleagues at Myriad were, quite possibly, even more upset. Soon after they discovered BRCA1, and the elation gave way to the hard realities of turning their find into a marketable product, the Myriad scientists knew they were in trouble. Myriad hoped to build a business as the country's premier gene-testing laboratory, providing the revenue and profits to pay back investors and fund new searches for genes underlying a wide range of other cancers and common ailments that lay hidden within the Utah genealogies. Having filed a patent for BRCA1, Myriad hoped to be the exclusive owner of the rights to commercialize the gene, as well as the genetic test to detect it, producing revenue from the test's sale or from licensing its use to others. But to meet that corporate goal, the Myriad researchers needed to create a simple process that was relatively inexpensive, completely reliable and available in a format that could be easily packaged and marketed to doctors, geneticists, and other appropriate health care providers.

"We took our responsibility as the patent-holder very seriously," Skolnick says. "We wanted to make the test as near-perfect as possible, because we knew its results would affect people so dramatically. We were building a business that would grow only if its product was reliable. If we screwed up this test we'd have a credibility problem that would taint our ability to provide other tests when we found other genes down the line."

The Myriad researchers faced significant problems. "The gene's size was simply brutal," Skolnick says. Its coding region, the part responsible for producing a protein, consisted of over 5,500 base pairs, making it one of the longest genes ever discovered. Immediately, Myriad reached out to researchers who had been collecting families, and together they began examining the lengthy gene in those family members, scanning for mutation sites. Myriad hoped to find a few spots common among the families where jumbled or misplaced DNA letters disrupted the gene. But instead, "we found a horror story," Skolnick says. "Maybe we shouldn't have been surprised."

Within a few months, scientists found that there were dozens of places where the gene could be mutated. In fact, almost every family studied carried its own distinctive mutation. That meant that each time Myriad or some other laboratory wanted to test an individual for the gene, the researchers would have to uncover the precise letter-by-letter sequence of the gene's more than 5,500 nucleotide units, a time-consuming and extremely tedious process even using new automated gene sequencing machines. Moreover, each sequence would have to be 100 percent error-free, because scientists found that even one letter out of place could produce a cancer-causing mutation. A mistake in sequencing would produce a result that suggested a defect when one didn't exist, or might overlook a mutation that was present.

By January 1995, about the time Stein first contacted her nearby medical center, "we realized we'd have to do full-gene sequencing for each and every test sample," Skolnick says. "Nobody had ever undertaken a sequencing project of that magnitude. No medical diagnostic test of this kind of complexity ever existed before." When Myriad found BRCA1, Skolnick had said he would have a commercial test available for wide use in two years, "and I was committed to that timetable, even though doing so was going to be very rough," he says

In the following months, Myriad undertook a massive scale-up of its operations. It moved to larger headquarters, tripled the size of the space occupied by its clinical laboratory operations, spent several million dollars buying 22 automated gene sequencing machines—apparently giving it more of the high-tech equipment than most pharmaceutical makers. Myriad's staff, which had grown from just 8 people in 1992 to 45 in 1994, doubled in 1995 to 98 employees, then doubled again in 1996 to almost 200 staffers, most of whom were brought on board to meet the new test's demands.

"We had to rewrite the [sequencers'] computer software, we had to create a system of tracking a [blood] sample from the moment it left the doctor's office until we processed it and sent back an answer, we had to reduce the amount of [test-tube] reactions by one-fifth, create a first-class quality-control system from scratch, and, of course, design a counseling program to help educate providers and test subjects as to precisely what a test result means," Skolnick says with an appropriate breathlessness. "Everything we were doing had to be invented because nobody had tried to commercialize a genetic test that would be this widely-used before."

Even while Myriad was gearing up its full-length sequencing operations, a few research medical centers, such as Memorial Sloan-

Kettering in New York and the University of Pennsylvania, sent samples to the company to detect mutations in living members of families in which the gene was believed to be causing cancer. Several
other research labs, such as the one approached by Rachel Stein,
also were conducting limited tests of women who had cancer in their
families. Once a specific mutation was found in an affected person,
the researchers felt comfortable looking for the same mutation in a
healthy relative who like Stein was as yet cancer-free.

But then scientists in Montreal, at Sloan-Kettering, and at the
National Cancer Institute made a surprising discovery. Kenneth
Offitt at Sloan-Kettering was collecting and cataloguing many
women who sought the gene test when it became available. Offitt
quickly struck a research agreement with Myriad, making him and
his patients among the first to use the new gene test. As Myriad
began scanning the genes of affected family members sent to it by
Offitt and other doctors, an unexpected finding emerged. In family after family, one mutation stood out as being more common
than any other.

"At first I thought this was very odd," Offitt recalls, "because
our experience had been that every family's mutation was different." But then Offitt took a closer look. Based in New York City,
and with a practice that drew heavily from the metropolitan area,
Offitt treated many women who were Jewish, and whose heritage
could be traced to the once-isolated eastern European Ashkenazi
Jewish community. (Ashkenazi describes the Jews who congregated in central Europe from the Middle Ages to the late 1800s.)

By early 1995, Stephen Narod, a Montreal oncologist with a
large practice favored by many in that city's Jewish community,
had a startling insight; many of the women he treated who had a
family-based cancer shared one mutation, even though they were
apparently unrelated to one another. At a spot in the gene, the
185th nucleotide in a region numbered 23, the scientists found
that just two of the gene's thousands of nucleotide base-pair units,
a single adenine (A) and a single guanine (G), were consistently
deleted, or missing. Called 185delAG, scientists soon realized this
mutation was the commonest of the more than 100 different mu-

tations uncovered in breast and ovarian cancer families. In a landmark study reported in April of 1995, researchers at the National Cancer Institute said that 8 of 858, or almost 1 percent of individuals of Ashkenazi Jewish ancestry whose blood previously had been collected for screening of cystic fibrosis and Tay-Sachs disease, also carried the 185delAG mutation in their BRCA1 gene.[5]

Several subsequent surveys found a similar 1 percent prevalence for the mutation among other pools of Ashkenazi Jews. Later, several research teams found that two other mutations, one called 6174delT in BRCA2 and another 5382insC, in BRCA1 were also commonly found in Jews tested, meaning, all told, that about 1 in 40 Ashkenazi Jews carried a mutation for BRCA1 or BRCA2. Scientists found that in the rest of the public, mutations to the genes were found in about 1 in 200 people.

"We were very much surprised that the actual carrier frequency of this mutation was so much higher than we expected," Sloan-Kettering's Offitt says. Scientists explain the high prevalence by pointing out that in Europe the Jewish community has remained insulated by cultural and religious practices that encourage marriages within the ethnic group. In other words, the American Jewish community retained many of the attributes of the Utah pioneers; a gene carried by an ancestral founder had been conserved throughout the generations, passed among descendants whose gene pool wasn't often diluted by an assimilation with other populations.

Indeed, Offitt and others suggest that Jewish history helps explain the gene's strong modern-day presence. At about the time of the Roman Empire, the world Jewish population was about 6 million to 8 million. But by the year 1100, Jewish numbers had dwindled to less than 2 million, following centuries of persecution. "Any particular trait in that group when it had gotten so small was certain to become more ingrained as the population grew more robust in later years; that's simple genetics," Offitt says. "The breast cancer gene is no different than any other genetic trait that was common in those years when the population declined."

At first scientists didn't know what to make of the discovery.

"I'm not sure how it can help my patients one way or another," Offitt said, noting that even though it was a commonly inherited gene variant among Jews, in the larger population, in which 211,000 new breast and ovarian cancers are identified each year, the role of this new information was unclear. Although he treated many Jewish women, "I couldn't presume their cancer was due to one of the three mutations," Offitt says. "And even if they tested positive, I wasn't certain yet what to do with the information."

Some Jewish groups were deeply disturbed that the gene would make some Jewish women believe they carried a problem more troublesome than other women did. "The truth is breast cancer is no more common among Jews than in the general population," Offitt says. "Although it may be also true that the inherited version of the disease may account for more breast cancer among Jews than among others."

During much of 1996, increased media attention about this unexpected discovery angered some women who believed women were being misled about the information. "It just makes me feel very uneasy, having Jews singled out this way," says Cathie Ragovin, a psychiatrist in Weston, Massachusetts, who is a breast cancer survivor and is active in cancer care advocacy groups. "There is an enormous concern about using this as another way to discriminate against people." She adds: "Women are anxious about this disease as it is. This business of a common gene among Jews, heightened their fears. It's a terrible thing to do to people."

Ragovin says that as news reports of the common gene variants spread into the Jewish community she and other activists received calls from "traumatized women, people frightened that they are carrying this gene because they are Jewish."

In New York and Boston, breast cancer activists in the Jewish community called for a moratorium on the aggressive commercial sale of the breast cancer test, worried that oncologists and surgeons not well-versed in the complexities of genetic risk would overpromote the benefits of getting a test result. "I'm worried that some Jewish women will think there is an epidemic due to this gene that they need to run out and do something about," says Judi

Hirschfeld-Bartek, a lawyer in Boston. "I'm not against the test; I'm worried about how it's being used."

But, for Rachel Stein and Staci Mishkin, the Ashkenazi gene variations looked to solve their gene-identification problem. A second cousin—the granddaughter of a woman who was the sister of Rachel's and Staci's grandmother—had had breast cancer twice and was getting tested for the gene. The second cousin's mother was alive and her blood was analyzed. Sure enough, the mother and the daughter carried the 185delAG mutation in their BRCA1 genes. "Once [the second cousin] tested positively, Staci and I realized we could finally be tested, too," Stein says.

Knowing that the Stein family mutation was the common Ashkenazi gene variant, researchers at the nearby medical center decided it would be relatively simple to examine Stein's DNA for the same mutation. Meantime, Mishkin had her blood analyzed for the same mutation at a company in Virginia called Genetics and IVF that was advertising to the Jewish community that it could quickly screen a woman's DNA for the presence of the three mutations common among the Ashkenazi descendants.

The news was great. Neither Stein nor Mishkin carried the mutation. "I thought, 'Wow! We don't have it,'" says Stein. "We were both ecstatic." But since the family's pedigree indicated that a gene was likely at fault, the researchers at the medical center where a tumor block from Stein's mother had been stored decided that to be safe they ought to check for the presence of either of the other two Ashkenazi mutations. Sure enough, DNA left over from Stein's mother carried a minute blemish in the BRCA1 gene, an extra cysteine (C) nucelotide at the gene's 5382 address. The lab workers then turned to Stein's blood again, this time scanning for the 5382insC mutation. In the summer of 1996, four months after being told that she had escaped the defect that caused cancer in the family, Stein was now advised that, in fact, her blood carried the same gene defect that presumably had killed her mother.

"I was upset, very very upset," says Stein. "But I wasn't surprised either." Stein called Mishkin and told her what happened. The researchers at Genetics and IVF returned to Mishkin's sample, too.

And as with Stein, Mishkin was found to carry the same mutation. "Even when they told me earlier that I didn't have the mutation, I still assumed I did," Mishkin says.

"The problem now was what were we going to do about the results," Mishkin said. "If [Rachel] hadn't pressed for it I don't think I would have been tested. But now I knew and there was no way to hide from the truth. My husband and I went back and forth, back and forth, for months about what we should do. It was terrible."

Mishkin and Stein now had to confront an issue that was tearing apart the medical community: should a woman who tests positive be told she should have her breast and ovaries removed. "To me there's no debate," says Mark Skolnick. "No matter how unappealing, the ovarian surgery is an important option." Skolnick and others believe that the surgery is the only significant approach to preventing ovarian cancer for women with the defective gene. Unlike breast cancer, often detected early by self-exam or periodic mammograms, ovarian cancer usually isn't found until it has advanced into a late and deadly stage, when it already has spread to other tissue. "A woman who removes her ovaries reduces her risk," Skolnick says, "and she likely saves her life."

But many health officials—geneticists and policy-makers—worry about doctors making this kind of blanket recommendation, no matter how well-meaning. Indeed, some charged that Skolnick was oversimplifying a very complicated medical and psychological situation because he wanted women and their doctors to believe that even learning they carried the gene defect could produce a positive health outcome. By late 1996, Myriad believed it had overcome many of the obstacles to commercial testing, and in October announced that it was selling the gene test: $2,400 was the cost for a full-gene sequence, and $350 to search for a mutation already identified as existing within a family.

"To be perfectly honest, Mark has a conflict of interest that is potentially very dangerous," says Neil Holtzman, a geneticist at Johns Hopkins School of Medicine. "Skolnick's group discovered BRCA1 and he and his group are understandably held in high esteem. I worry that doctors and medical providers will give his recommendations more weight than they should because of his well-deserved reputation as a scientist. But when it comes to using the test, and to making choices depending on its outcome, he has a company now that stands to profit handsomely from the test's use. People should understand he stands to benefit from a particular point of view."

As the new era of gene-testing began to emerge in the early 1990s, Holtzman had gained a reputation as being an especially thoughtful genetics watchdog. The title of his 1989 book, *Proceed with Caution,* aptly captured his point of view. In 1996, Francis Collins named Holtzman to head a special task force to review genetic testing, and the breast cancer gene test, in particular. Following months of hearings, Holtzman and many of his colleagues on the task force were struck by the fact that the gene test was entering wide use without having to meet any federal regulations. Test developers such as Myriad, OncorMed, or independent medical centers weren't required to meet any specific guidelines for quality control or for marketing.

"I believe the sale of these tests by commercial interests in particular should be regulated by the Food and Drug Administration," Holtzman says, noting that most clinical laboratories that offer a wide range of blood tests for doctors are strictly regulated. Moreover, companies that sell test kits to these laboratories to analyze such things as cholesterol or the presence of certain bacteria must have their tests approved by the FDA before they are allowed to sell them.

Especially worrisome to Holtzman and others, however, was the kind of information given to patients. "Few doctors truly understand that this test does not provide a simple yes-no kind of result," Holtzman says. "This is a test that provides patients and doctors with a risk factor result, and very few people understand how to evaluate or act on risks."

For instance, research in high-risk families showed that people with the gene had an 85 percent risk of developing breast cancer sometime in their life. But that figure arose from families where cancer is especially common. In other families, different mutations might actually confer a lower lifetime risk, perhaps closer to the 56 percent figure. Moreover, analysis of risk by age group shows that even in the cancer clustering families, a gene carrier's risk at age 25 is well below 10 percent, and that risk rises only to about 35 percent until age 40. Only at age 65 does the likelihood of cancer actually reach 85 percent, meaning that 85 percent of all carriers in the high-risk families developed cancer by age 65.

"But, right now, absent better experience, we are left telling patients they have a 56 percent to 85 percent risk, when in fact, depending upon their age and other factors, that probably is not their true risk at all," says Michael Kabak, the San Diego geneticist, who in 1996, as concerns about the test began to mount, was named to several research panels struggling to develop guidelines for doctors using the test. "We are not a society that understands probability well, how to interpret it or how to explain what it means to people, especially to someone who has just received some very bad news and wants advice—is really hoping for a recommendation—from a medical provider on what can be done."

"The few people who are trained in providing this kind of risk-assessment are genetic counselors," Kaback says. "But there are too few of them to handle the many people who are soon going to be demanding these tests."

He adds: "Right now, if we had a pill you could give someone who had the mutation, then I would be in favor of a national program encouraging every woman to be tested. But without any clear treatment, I think we need to be very cautious about what we say to people and how aggressively we allow people to promote the test's use."

Mary-Claire King says that she has come across women in high-risk families who carry the gene "and lived well into their seventies and eighties without ever getting cancer." King says this powerful evidence persuaded her that "simply promoting surgery as an answer to gene carriers was unfair and wrong." King argued strongly that anyone offering the test should be required also to offer extensive counseling. "Women need to be told compassionately that [having a mutation] in this gene is not a guaranteed death warrant."

By 1996, a formal protocol for advising those having the gene test was established at some of the country's more prestigious cancer research centers. At Columbia-Presbyterian Medical Center in New York, officials installed a full-time genetic counselor in an office just a few feet down the hall from where surgeons and oncologists see patients referred to the hospital for care. "The advent of this test has changed everything for us," says Karen Antmann, an oncologist who directs the hospital's breast cancer research program. "For the first time, we are aggressively concerned about family history even for a woman who already has cancer. That's something very new."

Antmann says, for instance, a woman with breast cancer and a family history who tests positive for carrying a gene mutation will be advised that there is a higher-than-usual chance of the cancer recurring in another breast or for a cancer to strike in the ovaries. Armed with the news that a patient was a gene carrier, the oncologists might want to recommend a more aggressive use of chemotherapy, radiation, or even prophylactic surgeries, Antmann says.

At Columbia-Presbyterian, a woman who is identified by the doctors as having a strong family history—at least one or two first-degree relatives with disease—and who is interested in information about the test is walked over to Donna Russo, the breast cancer program's own genetic counselor. After a session that can last an hour or more, the patient is asked to think about whether the test is desired. Even someone who understands the test's limits and benefits and wants the test is asked to return at a later date, when blood is drawn. When the results are ready several weeks later the patient and her family are offered still another lengthy session with the counselor and physician. Depending on the test's outcome, the patient then may be offered an appointment with a staff psychologist specializing in breast cancer issues. And, of course, the patient with a known mutation is advised that the only medical intervention, beyond monitoring, is preventive surgeries.

Antmann says that an informal survey found that of women in high-risk families offered the test, about one-third wanted it right away, about one-third said they wanted to think it over, and one-third declined to take it. Of the one-third that put off the decision, most don't want it, Antmann says. "This is not something everyone wants," Antmann says. "If we're doing the counseling right, people will feel comfortable saying 'no.' "

But observers worry that as soon as a medical provider brings up the option of a test and of preventive surgery, many patients will embrace action more strongly than they ought to, especially given the current state of medical knowledge. "People come to a doctor seeking treatment advice, but a positive result in this test does not allow a doctor to recommend treatment," says Arthur Caplan, one of the country's foremost medical ethicists, who runs the University of Pennsylvania Center for Bioethics. "A woman who is told she has a mutation for the breast cancer gene can be advised of the options, but the doctor must not make a recommendation one way or the other. This is very hard. But any doctor who tells a patient that surgery is better than close monitoring is giving advice right now without any research to back the claim. This is simply unethical and wrong. There may come a time when research accumulated by following these women over many years will point which way to go. Meanwhile we are doing the tests, despite our limited knowledge."

Caplan also worries that even when providers are careful, the new genetic tests can produce consequences no doctor or genetic counselor can foresee or ever be fully prepared to handle. One day, the counselors at the university hospital called him over for a consultation. A woman was tested positive for a gene mutation and her husband wanted their 14-year-old daughter tested, too. The university's policy was not to inform a woman until she was at 18, and Caplan told the father that it would be unethical to test the daughter. "The fellow stalked out, angry at us for denying him the information he wanted and he went and had the test done elsewhere," Caplan says. Later, Caplan learned, when the girl was found to be positive for the mutation carried by her mother, the family had the girl's budding breast tissue removed.

"You know, the Hippocratic oath says to doctors, 'First, do no harm,'" Caplan says. "Is it possible that this information violated that oath?"

Indeed, many doctors and geneticists say that until more precise data arise from research analyzing the mutations and their impact on people, they are caught in a deeply troubling period. "Right now, we can't tell a patient that one type of mutation is more likely to produce cancer than another mutation," says Jill Stopfer, a genetics counselor specializing in genetics at the University of Pennsylvania. "We can't tell them if their particular mutation will cause breast or ovarian cancer, whether the mutation they have correlates with recurring cancer or not. We get bombarded with questions and have darn few answers. Our job is to lay out the unknowns, and hope the patient's decision is as informed as possible."

Rachel Stein and Staci Mishkin entered this minefield the day they tested positive for the BRCA1 mutation. "It is our experience that women who have lived with breast cancer in their family, who have seen close up what it can do to its victims and to close relatives, are the ones who are more likely to seriously consider prophylactic surgery," says Barb Weber, who had moved her genetics research laboratory to the University of Pennsylvania. "For them the surgery is not nearly as grisly an idea as it is for others."

And so, in January 1997, Mishkin had both breasts and her ovaries removed. Mishkin says that although it took her many months to decide on that course of treatment, in the end her choice "made perfect sense." She already had two children so she felt somewhat less troubled about losing her ovaries. "I was upset about becoming postmenopausal, about what it would mean to lose my hormones, at my age," she says.

Still, "I believed that removing my breasts and ovaries would lower my risk," she says. "They told me I could do something to lower my risk and I wanted to do that." But even so, she concedes, she is startled by the turn of events. "If [Rachel] hadn't pressed us to get tested, I wouldn't have done it. I would have lived with the uncertainty. But once I knew I had the gene, I had to do something about it. I couldn't make believe I wasn't at risk."

Some ethicists argue that, in fact, the test may well have taken away Mishkin's inclination to be uninformed. "One of the most troubling things about this test, as compared to other medical tests, is that a positive result doesn't just affect you, the way, say, most other tests do," says Holtzman of Johns Hopkins. "As soon as we inform someone [she is] positive, we are imposing that news on the entire family. Very few providers have the resources or even the interest to help other members of the family deal with this."

More than two years after first inquiring about the test, Stein also decided to undergo surgery, scheduling hers two months after Mishkin. "I have absolutely no ambivalence about this," she said a few months prior to the operation. "Unlike my mother and my aunts, I can do something. It's a very good feeling. It's a miracle. Maybe the discovery of this gene will someday lead to a pill, and by the time my daughter is my age she won't have to face this. I have to believe she will have choices I don't have. But for me, right now, this feels okay."

Indeed, many women who seek the test say they don't expect a foolproof guarantee from the health care profession, but simply information about the unknown upon which they can act if they choose. Geneticists and cancer doctors say that even some women who have breast cancer and receive news that they have a mutation seem to benefit emotionally from getting the test results. Ken Offitt says just about every day he sees someone in whom cancer appears to run in the family. "We really have to evaluate each person individually to help us decide whether the test makes sense, whether it can provide some clinical or emotional benefit," he says.

Offitt says despite his growing experience using the test, he is constantly surprised by its power, by the unpredictable impact it has on some people's identity and sense of well-being. One day, for instance, he met with a woman in her 60s who had breast cancer years before. Offitt was seeing her because she was now being treated for advanced ovarian cancer. "Here she was undergoing very aggressive chemotherapy, she had lost most her hair and had a catheter protruding from her [gut], all of which is quite humiliating," he says. "She was going through a very tough time, and then on top of that she tells me she wants to get tested for the gene."

Offitt was confused, at first, and believed the woman had misunderstood the test's usefulness. Perhaps she mistakenly believed it could somehow help her treatment. The woman didn't have any children and had lost her sisters in the Holocaust, so nobody in her family could benefit from the test results. "She was very articulate, a very strong person," Offitt says. "As we talked she told me that she wanted to know if she had the Jewish mutation. She told me she had worked hard to be healthy her whole life, she had been a good person, had been careful about her diet and had exercised. And yet here she had this terrible health problem. She wanted to know if she had the gene because she wanted some explanation for the madness that had happened to her. She said it would mean the cancer was not her fault, that it was something that she had nothing to do with."

Offitt says, "It was a very poignant moment for me. When I was beginning my practice in oncology in the late 1970s, we had no clue as to what caused cancer. Maybe it was environmental; there was all sorts of crazy guessing going on. For many patients there was this sense that cancer was this malevolent thing that came from God knows where. Well, for some patients we know where it comes from. And for some of them, this has a healing, almost restorative effect. It can be empowering, just knowing. And right now, maybe, that's the best we can hope for."

Epilogue

In early 1997, a research report in *Science* hinted that still another lethal cancer was surrendering its secrets to the relentless probing of gene-hunting scientists. In the report, researchers said they had found the first crucial clue to the biochemical origins of a common type of brain cancer, a particularly hard-to-treat killer whose victims each year in America number about 11,000. After many futile years of searching, the researchers had finally latched onto a scientific lead that, if pursued vigorously, would likely produce a strategy for attacking the disease.

"It's a piece of the puzzle," says Ramon Parsons, a protégé of Bert Vogelstein, who made the discovery at his lab at Columbia University School of Medicine in New York, where he had gone after "graduating" from Vogelstein's team at Johns Hopkins. "Before this we knew practically nothing about this cancer, how it arises, and how we might go about finding a treatment."

Parsons found a gene that was consistently mutated in tumor cells removed from patients with the brain cancer. To Parsons's trained eye, honed during four years at Vogelstein's lab, the hitherto unknown gene appeared to be a new and important tumor suppressor. When the gene called PTEN is mutated and unable to carry out its job, brain cells normally under its care are allowed to run rampant.

But by 1997, more than eight years after Bert Vogelstein first found the tumor suppressor role of p53, uncovering still another tumor suppressor, despite its considerable importance to brain cancer research, was seen by many not as a breakthrough, but instead as validation of the emerging genetic model for cancer genesis. To those in the field, the research confirmed that a new, fundamental theory of cancer was now in place.

"Most every significant basic research project we fund is seeking to characterize the genes that turn tumor growth on and off," says Richard Klausner, the director of the federal government's National Cancer Institute. "That represents a major shift in the direction of cancer research over the past few years." Klausner adds: "It's becoming pretty clear that the 1990s will be seen as a time when scientists finally uncovered the strategies that will be the foundation for every future research effort to eventually conquer cancer." Each of these strategies in some way involves understanding the genetic basis of a cancer.

Indeed, in the months preceding Ramon Parson's PTEN gene discovery, several other major findings, taken together, suggested that new gene-based insights already were yielding significant advances in detection, prevention, and treatment of cancer:

More than two years after BRCA1 was isolated, researchers finally began piecing together how a mutation to BRCA1 can trigger breast cancer, suggesting potential avenues for fighting the cancer with drugs.

Scientists produced a reliable test kit doctors can use in detecting the recurrence of a bladder cancer in patients by identifying p53 mutations in urine, thereby commercializing the experimental test first used by Johns Hopkins researchers to detect evidence of Hubert Humphrey's lethal cancer.

Researchers in California produced a computerized "biochip" that appears able to detect minute mutations in BRCA1 and p53, suggesting that doctors may someday soon be able to detect cancers earlier or identify a person's inherited susceptibility to cancer with tests that are simpler and less expensive than previously produced.

Doctors began testing some tumor cells for the presence of p53 mutations, using the results to guide which of several types of chemotherapy would work best against an individual patient's cancer. But perhaps the most notable developments emanating from the research labs about the same time as Parsons's discovery were the first hints that prostate cancer, much like breast and colon cancer, can be caused by the inheritance of a mutant gene. If the preliminary gene findings regarding prostate cancer that surfaced in late 1996 and early 1997 are confirmed as expected, scientists believe that most all basic research efforts in the future against the disease will be aimed at elaborating the genetic pathway that transforms a normal prostate gland cell into a cancerous tumor. Indeed, the breakthrough resulted in part from a rising concern about the growing prevalence of the cancer. But it also resulted from a fierce scientific race, much like the ones in breast and colon cancer, among some of the country's foremost gene-hunting labs to track down the genetic cause of the disease.

In recent years, prostate cancer has begun striking Americans with an alarmingly higher incidence, growing by 65 percent between 1980 and 1990 and by 200 percent in the years after that as many men lived longer and the disease was diagnosed more frequently. In 1996, about 340,000 Americans developed prostate cancer and about 41,000 men died of the disease—only a bit less than the number of women who died that year of breast cancer—making it the second leading cancer killer in American men after lung cancer. Numerous celebrities, such as General Norman Schwarzkopf, rock-and-roll star Frank Zappa, the famed investor Michael Milken, golfing great Arnold Palmer, and former Senator Robert Dole have contracted the disease in recent years, raising the public's awareness.

"For men, prostate cancer is becoming the equivalent of what breast cancer has been for women, an increasingly common problem," says Donald Coffey, a cancer specialist and researcher at Johns Hopkins University. "Until now nobody knew how it arose, what its molecular origins were, and what we could do to prevent or treat it. Many of us believe, based on what we have now learned from studying other tumor types in the past few years, that genetics holds the key for solving this growing problem."

For many years, even as researchers uncovered the genetics of several common cancers, most cancer specialists scoffed at the notion that prostate cancer involved the defect to a single gene that could be inherited in a mutated form. In 1982, Lisa Canon-Albright, a researcher working with Mark Skolnick at the University of Utah—using Skolnick's famed computer database and the Utah Cancer Registry—reported the discovery of several extensive Utah families in which prostate cancer appeared to be inherited.[1]

"Most of us believed this was a rarity, something isolated to a group of families in Utah," says Patrick Walsh, a noted prostate cancer surgeon who directs Johns Hopkins's Brady Urological Institute in Baltimore. Walsh and his colleagues in the field didn't consider the disease to be like colon cancer or breast cancer, in which it was commonly known that having a sibling or parent with the disease was a risk factor. "I remember hearing about the Utah families, being intrigued, but not giving it too much thought beyond that," Walsh says.

But through a series of unexpected events, Walsh was soon drawn intimately into the gene-hunting fray. It began with a surgical advance that made Walsh world-famous. For many years, prostate cancer patients avoided what many doctors believed was the best therapy, surgical excision of the gland, an inch-long, walnut-shaped organ. That's because surgery meant the end of a healthy sex life for most men.

The prostate sits just under a man's bladder; its main job is producing the semen released during orgasm. The male urethra, the canal which carries urine, passes from the bladder, through the prostate, and then into the penis. In most men as they age, the prostate enlarges, a process tied to the release of the same male hormones that cause age-related baldness. As the prostate increases in size it puts pressure on the urethra, making it difficult for men to urinate completely, and often causing a man to awaken frequently during the night in order to relieve urine left in the bladder. This common enlarged prostate condition is often treated with drugs, and in severe cases, with surgery, but often an affected man will simply put up with the annoyance since both drugs and surgery have side effects.

Much more serious, however, is a problem that occurs when a swelling of the prostate is caused by a tumor expanding within the gland. The early symptoms of cancer are similar to the benign condition, and often the cancer never becomes life-threatening, confined as it is to the gland. But in many patients a fast-growing tumor can be dangerous, for its cells can escape the gland, spreading with murderous intent into the bladder, kidneys, or liver. In recent years, researchers have struggled to find a diagnostic test to detect a fast-growing prostate that would indicate the presence of a cancer. An increase in a substance released into the bloodstream from the gland, called PSA, can be a signal of growth, although a test for it is controversial because the test often can't identify whether the enlargement of the gland is the common benign sort or the dangerous cancerous kind. Doctors recommend that most men over 50 have an annual physical exam that checks the prostate to see if it is growing larger than is believed safe.

For years, Walsh was frustrated with treating patients who ignored or postponed therapy, thereby allowing their cancers to advance. Most patients deferred treatment because the surgery to remove the gland not only eliminated the production of semen, it also damaged or removed nerves that stimulate orgasm and penile erection. By 1984, after years of study and experiment, Walsh perfected a so-called nerve-sparing surgical technique. As word spread of his surgical invention, Walsh began attracting patients the world over. And he began seeing an increasing number of younger prostate cancer patients for whom salvaging an intact libido was especially desirable.

In 1986, Walsh treated a patient that made him reconsider the Utah study. The 49-year-old man sought out Walsh for his "nerve-sparing" surgery, but he also relayed some information Walsh hadn't heard before. The man's family was riddled with prostate cancer. His father, his father's three brothers and his grandfather had all died of the disease. Walsh was stunned.

"Here I was chairman of urology at Johns Hopkins and I simply did not know whether having a close relative with the disease raised your risk or not," says Walsh, a tall, thin and balding man whose usual reserved and formal manner becomes increasingly animated whenever he relates the moment that changed his career. "I decided then to characterize that risk," he says.

Over the next four years, as Walsh's surgical reputation grew and he attracted a huge patient following, he painstakingly recorded his patients' family histories. By 1990, with the assistance of several colleagues, Walsh had amassed an impressive collection of pedigrees on 691 men with prostate cancer, as well as family trees of the men's spouses. By comparing prostate cancer in the families, the researchers found that 15 percent of the men with cancer, but only 8 percent of their spouses, had a close relative with the disease. By analyzing the data, they found that men with a father or brother with the disease were twice as likely to develop the cancer as compared to someone without a close relative with the disease. Moreover, men with two first-degree relatives affected had a fivefold increase in risk of contracting the disease, while those with three affected relatives had a risk 11 times greater than someone without a relative with the disease.

"The data were astounding to me," says Walsh. "It meant that what the folks had seen in Utah was not unusual but indicated that prostate cancer had a strong family connection." In looking back, Walsh and others realize that if not for his nerve-sparing operation which attracted younger men to his practice, the urologist might not have been able to detect this familial link. That's because most men develop prostate cancer late in life, an age when many siblings or parents have likely died of other causes. Moreover, since the disease is relatively common, linking the disease to an inherited cause was difficult—a cluster in a family might simply be due to coincidence.

But Walsh's research suggested that an inborn susceptibility to the cancer existed. Inherited prostate cancer was rare, but, he calculated, it likely accounted for 10 percent of all prostate cancers. Given the rising incidence of the disease, uncovering the cause of one out of ten cases would certainly be a major accomplishment. After recruiting the assistance of geneticists at Johns Hopkins, an analysis of the pedigrees he had collected suggested that the disease was indeed being caused by the passage in the families of a single, dominant-acting gene.

"By 1992, we had a very clear model upon which to base our research," Walsh says. "Mary-Claire King had shown that if you could collect enough large families with excessive amounts of cancer, this gene, if it existed at all, could be uncovered. We decided if we were going to make headway we needed to gather as many of those families as we could."

In 1993, Walsh gave the keynote speech at a scientific meeting of the American Urological Association in which he described his efforts to track down the genetics of prostate cancer. After several doctors in attendance came up to him with anecdotes about one or two cases they treated where the cancer might have a family link, Walsh decided to mail out a letter to 8,000 colleagues in the United States asking them for help. Soon Walsh was awash in responses. By this time, he was being assisted by William "Billy" Isaacs, a molecular biologist in the urology department, and by the biologist's sister-in-law, Sally Isaacs.

"Sally began collecting and analyzing the patients being referred to us, picking out those that looked as if they had the highest likelihood of involving a gene," says Walsh. "It was her job to track down the patients, their family members, and get the blood and DNA from them." Meanwhile, Billy Isaacs's lab began sifting through the genetics of the families collected.

"Even with the families coming in from Pat's mailing, we knew we needed more and larger families if we were going to find a gene," says Billy Isaacs.

Then in early 1995, two developments spurred the research. As news spread that Walsh had shown that prostate cancer had a familial association and that he was beginning to amass an impressive array of families, several gene-hunting scientists began courting Walsh and Isaacs, offering the expertise of their molecular biology labs in return for a 50-50 collaboration. Among those who came calling was Francis Collins.

After having lost the race to clone BRCA1, Collins's lab at the National Center for Human Genome Research, run by Jeffrey Trent, was seeking a new high-profile gene to attack. Following an introduction by Bert Vogelstein, a mutual professional acquaintance, Collins and Trent told Walsh that they could use their gene probe technology to sift through the families' DNA to map the location of a gene, if it existed.

"Walsh had collected a very impressive set of families," says Trent, "but we told him and Isaacs that the odds would be increased with even better families still."

The solution didn't take long to arrive. Soon after the collaboration was agreed to, Walsh's effort to find families was described in a brief article in *Parade* magazine, the Sunday newspaper filler with a circulation of about five million readers. Walsh also appeared at about the same time on the *Phil Donahue Show*. Within days of the publication and television appearance, the Hopkins researchers were flooded with telephone inquires, eventually generating responses from about 2,500 families within several weeks. Picking through them, Sally Isaacs identified 91 families—79 in the United States and 12 from overseas—that met the criteria of having at least two or three closely related family members with the disease. By early 1996, Trent's lab had enough family DNA—samples from over 600 family members—to begin a search throughout the genome for markers that were linked closely to the gene presumed to be involved in the inherited disorder.

"We worked quickly, but in the end we tested over 350 markers spread at random throughout the genome," Trent says. In the summer of 1996, the Genome Center lab found a DNA marker in a small spot of chromosome 1 that was consistently linked to people with disease, but rarely so with those family members unaffected by the cancer. "The discovery was a testament to the extraordinary advances in molecular genetics," says the National Cancer Institute's director, Richard Klausner. "The swiftness of Collins's lab really shows the kind of progress being made in the laboratory."

In late November, just before Thanksgiving of 1996, the researchers announced their find in the journal *Science*, and the scientific and lay media throughout the world hailed the discovery.[2] Dubbing the gene HPC-1, for human prostate cancer-1, Walsh and his colleagues predicted the gene hidden somewhere in the chromosome 1 region was likely to account for about one-third of the 10 percent of all prostate cancer cases that are inherited. "I am profoundly excited by what this may mean to the people I treat, to their families, and to the many people who carry this gene and aren't yet aware of it," Walsh says. "The research into the genetics of this disease is what gets me up out of bed and going every day."

Even before the gene's location was announced, researchers working for Mark Skolnick at the University of Utah and at Myriad Genetics began scanning through several dozen families they had collected over the years to try to confirm the chromosome 1 linkage. Another team of scientists at the University of Washington, largely funded by research grants from the controversial investor, Michael Milken, also began pointing their molecular biology weapons at the same site. Says Skolnick: "We are taking this effort [to find the prostate cancer gene] just as seriously as we took the effort to find BRCA1. This is a race and we want to win."

In an interview with the *Los Angeles Times*, Francis Collins said: "When we began this project, it was like Columbus setting off across the Atlantic Ocean in boats and not knowing if he was going to find land or drop off the edge of the world. Now we have land in sight."[3]

The excitement, researchers say, was justified. "What these guys proved is that there is a strong genetic component to a very common cancer among men," Bert Vogelstein says. "Now we surely suspected that was true. And it's certain that this won't be the only gene involved in the cancer. It's very likely there will be other genes that can be inherited and still other genes involved in the progression of the disease that will be uncovered by studying

DNA alterations within tumors. But what Walsh and Collins accomplished was a start, and that's critical these days."

In Seattle, where she was continuing her studies of the breast cancer genes at the University of Washington, Mary-Claire King greeted the prostate cancer gene find with tempered and philosophical enthusiasm. "These boys are now in a very stiff race" to find the actual gene, she says. "We all believe these gene hunts are going to change the direction of cancer research forever. That's why we do what we do. Ultimately our goal is to cure these diseases. What's so great is that these days we have the technology at hand and a pretty clear idea about how to go about trying. Do I think all this someday will lead to a cure? You bet I do."

Acknowledgments

This book grew out of my daily coverage at *The Wall Street Journal* of the sweeping changes that have transformed health care in America during the past two decades. In that time I witnessed and often wrote about two revolutionary movements: the tumultuous and explosive growth in the delivery and cost of health care services, and the remarkable, sometimes seemingly miraculous, advances taking place in medical science. For many years my job was to write about the intersection of these developments: how new technology spurred the growing use and cost of care, and how the abundance of health care dollars was responsible in part for motivating corporate and academic scientists to produce new ways to attack disease.

I have had a unique opportunity, therefore, to describe the forces driving the surge in costs and medical knowledge. As a result, sometime in the late 1980s it became clear that the ability of scientists to decipher the makeup of human DNA would most likely have the greatest impact in shaping American health care in the early years of the next century. In 1990, I joined with my long-time colleague at *The Wall Street Journal, the* veteran science writer Jerry Bishop, to write a book, *Genome,* in an effort to describe the significance of these genetic advances. That book told of the discoveries that were heralding a new age in human biology, the era of gene-hunting.

Soon after *Genome* was published, it became clear these new gene discoveries were helping to unravel finally the mysteries of cancer. For several years I wrote numerous articles for *The Wall Street Journal* chronicling individual cancer gene advances. This book arose from a desire to describe how and why these new insights into cancer have occurred, and what effect they will likely have in the fight against the disease.

In tackling this new assignment, I offer much appreciation to my former co-author, Jerry Bishop, as dear and durable a friend as anyone could want. During the forty-two years he worked at the *Journal*—he retired in early 1997—Jerry developed a unique skill for reporting and writing about the drama, conflict, and exhilaration experienced by ordinary people who do extraordinary things while expanding the boundaries of knowledge. Through example and by the daily grind of editing and rewriting and advising, Jerry tried to pass along to me his special understanding that the scientific endeavor is a rich source of human sagas. I will forever be in his debt.

Indeed, I owe much to *The Wall Street Journal*, to the institution itself, and to my friends and colleagues there who make it an uncommon and challenging place to work. This book would never have happened without the initial backing of my editor for more than ten years, Neil Ulman. Neil is naturally curious about how complicated things work and especial adept at conveying life's complexities through compact and graceful narratives. I hope this book benefited from some of the skills he tried to impart.

I also want to thank the *Journal*'s managing editor, Paul Steiger; deputy managing editor Dan Hertzberg; national news editor, Cathy Panagoulias; and my current editor, Dennis Kneale, for allowing me to pause from my daily responsibilities so that I could report and write this book. Cathy, especially, cheered me on whenever I faltered, urging me to tell the *Journal*'s readers some of the many stories that served as the basis for this book. Thanks, also, to the encouragement of Dick Tofel, Barney Calame, and Dan Kelly on the *Journal*'s "9th floor."

Most especially, I could not have taken off the time for this book if not for the generosity of my fellow health care reporters Ron Winslow, Bob Langreth, George Anders, and Elyse Tanouye. Only rarely did they complain that my absence increased their workload. Jeff Tannenbaum also gets a special thanks for reading early drafts of the book and providing impeccable editing comments.

In this book, as elsewhere, I have tried to describe the human side of the science. That has involved intensive and lengthy reporting of many members of families consumed by cancer who agreed to provide scientists the raw material for the gene-hunting research. Telling these stories also meant that the scientific investigators themselves had to be willing to share with me the minutiae of their lives, through which I hoped to convey their curiosity, their ambition for fortune and fame, as well as their honest and often immeasurable desire to change the world. I was very fortunate that they allowed me to interrupt their work and intrude in their lives.

In particular, this book would never have occurred without the gracious participation of Mary-Claire King, Bert Vogelstein, Francis Collins, and Mark Skolnick, each of whom spent untold hours helping me to understand the science, the scientific process, and the realities of conducting leading edge research.

At Simon & Schuster, I owe much to my editor Bob Bender. Bob showed much compassion in helping me weather several storms in pulling this book together, and his experience and skill in organizing complex stories is reflected throughout the manuscript. Thanks also to the frequent support and patience of Johanna Li, and to the copy-editing of Philip James.

Finally, to my family, whose abundant love is a continuing source of strength and joy: thank you.

Notes

1: A MYSTERY SOLVED

1. Ralph H. Hruban et al., "Brief Report: Molecular Biology and the Early Detection of Carcinoma of the Bladder—the Case of Hubert H. Humphrey," *New England Journal of Medicine* (330:81, 1994), 1276–1278.

2. Edgar Berman, *Hubert: The Triumph and Tragedy of the Humphrey I Knew* (New York: G. P. Putnam's Sons, 1979), 273.

3. Ibid.

4. David Sidransky et al., "Identification of Ras Oncogene Mutations in the Stool of Patients with Curable Colorectal Tumors," *Science* (256:1992), 102–105.

5. Hruban, "Brief Report," 1277.

6. Ibid., 1277.

7. *Cancer Researcher Weekly* (Birmingham, Ala.: C. W. Henderson, May 16, 1994).

8. Berman, *Hubert,* 70.

2: SCIENCE FICTION

1. The family members described here have asked that their true identity not be revealed in order that their privacy be protected as a result of the sensitive nature of their experience.

Some details of the family's experience were first reported in a front page *Wall Street Journal* article in late 1992, only a few months after the events of that summer. For reference see Michael Waldholz, "Stalking a Killer: Scientists Near End of Race to Discover a Breast Cancer Gene," *Wall Street Journal* (December 11, 1992), A1. At that time, only the two sisters in the family agreed to be interviewed and to share a small portion of their experience with the *Wall Street Journal* readers. That was the first time the sisters were identified publicly with the pseudonyms "Susan" and "Janet." Following publication of the *Wall Street Journal* article, the two sisters agreed to retell the same small portion of their dramatic story in several print and broadcast media outlets. Since then, "Susan" and "Janet" have appeared on national television, and they have spoken at scientific and medical forums on the condition that they be identified with the names first provided in the *Wall Street Journal*. (On TV, they also appeared in shadowed silhouettes.)

In the time since, I have spent many hours interviewing "Janet" and "Susan," visiting with them and many members of their extended family in their homes and at other locations. Each of the family members agreed to be interviewed with the understanding that their real names would not be revealed. I have produced pseudonyms for them all.

3: "WELCOME TO CHROMOSOME 17"

1. The full discovery was communicated to the science community as the cover story in the journal *Science* on December 21, 1990. See: Jeff Hall et al., "Linkage of Early-Onset Familial Breast Cancer to Chromosome 17q21," *Science* (250:1990), 1684–1689.

2. James Watson retained his position at Cold Spring Harbor Laboratory. He traveled to Washington, D.C., a few days a week until he left the federal job in early 1993.

3. Watson interview.

4. Steve Narod et al. "Familial Breast-Ovarian Cancer Locus on Chromosome 17q12-23," *Lancet* (338:1991), 83.

5: TRICK OR TREAT

1. Francis Collins, "Medical and Ethical Consequences of the Human Genome Project," *Journal of Clinical Ethics* (2:260, 1991), 267; *Guidelines for Predictive Testing for Huntington's Disease* (New York: The Huntington's Disease Society of America, 1989).

2. Barbara Biesecker et al., "Genetic Counseling for Families with Inherited Susceptibility to Breast and Ovarian Cancer," *Journal of the American Medical Association* (269:1993), 1970.

3. Ibid., 1970

4. Several weeks earlier, I met with Francis Collins while attending a journalism symposium on genetics sponsored by the University of Michigan. During a private conversation regarding efforts to find the breast cancer gene, Collins told me about Susan's experience the previous sum-

mer. He also told me the rest of the family was to receive its test results at a mass counseling session later in the month. I told him I wanted to write about the family's experience for *The Wall Street Journal*. He said he would ask the family for permission, which he did during the session on Halloween weekend. I subsequently did write a front-page article about it. For reference see Waldholz, "Stalking a Killer."

6: THE GAME

1. The other applicant, the one who got the job, was Francis Collins.

2. Dan Fagin, "High Stakes Hunt to Find a Gene," *Newsday* (November 16, 1993), 76.

3. In 1995, Steve Friend moved to Seattle where he was affiliated with the University of Washington and the Fred Hutchinson Cancer Research Center.

4. "Mary-Claire King—For Bringing Humanism to Science," *Glamour* (December, 1993).

7: CANCER FAMILIES

1. Personal letter to Gary Wilson from Mary-Claire King, July 24, 1991.

2. Ibid.

3. Mary-Claire King and Allan C. Wilson, "Evolution at Two Levels in Humans and Chimpanzees," *Science* (189:1975), 107–116.

4. "Evolutionary Thinking," *Newsweek* (July 7, 1975), 40–41.

5. P. R. Broca, *Trait des Tumours* (Paris: P. Asselin, 1866–1869), Vols. I & II.

6. Gregor Mendel, an Austrian monk, first described the laws of inheritance in a largely unnoticed publication dated, coincidentally, in 1866. In a series of experiments with flowering pea plants, Mendel showed that specific traits of parent plants can be transmitted to offspring plants under an identifiable set of rules, which we know as Mendelian laws of inheritance. Under these laws, Mendel proposed that certain traits in his plants, such as a particular flower color, are dominant and others are recessive. That is, when two plants are cross-bred, if just one of the plants contains a dominant trait, half the offspring on average will express that trait. But in order for some offspring to express a recessive trait, it must receive the trait from both parents, not just one. The point here is that despite his intelligence, Broca had no way of knowing for certain that a trait, such as a vulnerability to cancer, could be passed from parent to child. The importance of Mendel's ideas was not recog-

nized until after his death in about 1900 when his written work was discovered by three scientists, after each had independently obtained the same kind of results.

7. O. Jacobsen, *Heredity in Breast Cancer: A Genetic and Clinical Study of Two Hundred Probands* (London: Lewis, 1946), 366.

8. Several states in the U.S. require doctors to submit names for similar types of cancer databases. One registry in Utah has become one of the world's greatest resources for ferreting out cancer families. It plays a crucial role in the breast cancer story and will be further discussed in later chapters.

9. L. S. Penrose et al., "A Genetical Study of Human Mammary Cancer," *Annals of Eugenics* (14:1948), 234–266.

10. Henry T. Lynch et al., "Hereditary Factors in Cancer," *Archives of Internal Medicine* (117:1966), 206–212.

11. Henry T. Lynch et al., "Tumor Variation in Families with Breast Cancer," *Journal of the American Medical Association* (222:1972), 1631.

8: GOLD MINE

1. Corwin Q. Edwards, Mark Skolnick, and James P. Kushner, "Hereditary Hemochromatosis: Contributions of Genetic Analyses," *Progress in Hematology* (New York: Grune & Stratton, 1981), 43–71. George E. Cartwright, "Hereditary Hemochromatosis," *New England Journal of Medicine* (301:1979), 175–79.

9: GUARDIAN OF THE GENOME

1. "Hottest Scientist of 1993? It's Bert Vogelstein," *ScienceWatch*, December 1993.

2. Alfred G. Knudsen, "Mutation and Cancer: Statistical Study of Retinoblastoma," *Proceedings of the National Academy of Sciences* (68:1971), 820–823.

3. Eric Fearon et al., "Clonal Analysis of Human Colorectal Tumors," *Science* (283:1987), 193–197.

4. Bert Vogelstein et al., "Genetic Alterations During Colorectal-Tumor Development," *New England Journal of Medicine* (319:1988), 525–532.

5. Peter Nowell, "Molecular Events in Tumor Development," *New England Journal of Medicine* (319:1988), 575–577.

6. Ibid.

7. Robert A. Weinberg, *Racing to the Beginning of the Road* (New York: Harmony Books, 1996), 50.

8. David Lane later moved to the University of Dundee in Scotland.

10: ISHMAEL'S TALE

1. Jean Marx, "New Colon Cancer Gene Discovered," *Science* (260:1993), 751.

2. Natalie Angier, "Scientists Isolate Gene That Causes Cancer of Colon," *New York Times* (December 3, 1993), A1.

3. Jukka-Pekka Mecklin et al., "Screening for Colorectal Carcinoma in Cancer Family Syndrome Kindreds," *Scandinavian Journal of Gastroenterology* (22:1987), 449.

4. Albert de la Chapelle, "Disease Gene Mapping in Isolated Human Populations: The Example of Finland," *Journal of Medical Genetics* (30:1993), 857–865.

5. Albert de la Chapelle, "Close Linkage to Chromosome 3p and Conservation of Ancestral Founding Haplotype in Hereditary Nonpolyposis Colorectal Cancer Families," *Proceedings of the National Academy of Science*" (91:1994), 41–51.

6. Mr. Mast's first name is Ishmael, but his surname and those of his relations are changed to protect their privacy.

7. Jane Green, "Development of a Screening Program for Hereditary Non-Polyposis Colon Cancer," *Development, Implementation and Evaluation of Clinical and Genetic Screening Programs for Hereditary Tumor Syndromes*, Ph.D. thesis, Memorial University of Newfoundland, St. John's: 1995, Chapter Five.

8. P. Peltomaki et al., "Genetic Mapping of Locus Predisposing to Human Colorectal Cancer," *Science* (260:1993), 810–812.

9. Lauri Aaltonen, "Clues to the Pathogenesis of Familial Colorectal Cancer," *Science* (260:1993), 812–816.

10. Marx, "New Colon Cancer Gene Discovered," 751.

11: CLONE BY PHONE

1. P. Peltomaki et al., "Genetic Mapping of Locus Predisposing to Human Colorectal Cancer," *Science* (260:1993) 810–812.; Lauri Aaltonen, "Clues to the Pathogenesis of Familial Colorectal Cancer," *Science* (260:1993), 812–816.

2. A-L. Lu et al., "Methyl-Directed Repair of DNA Base-Pair Mismatches in Vitro," *Proceedings of the National Academy of Science* (80:1983), 4639–4643.

3. Robert A. Reenan and Richard D. Kolodner, "Isolation and Characterization of Two *Saccharomyces cerevisia* Genes Encoding Homologs of the Bacteria HexA and MutS Mismatch Repair Proteins," *Genetics* (132:1992), 963–973.

4. Manuel Perucho and his colleagues at the California Institute of

Biology at San Diego are widely credited with first seeing the phenomenon of genetic instability in colon cancer cells. In late 1992, Perucho and
his lab identified alterations in certain nucleotide base pairs in colon
cancer cells. Perucho's research, however, wasn't published until June of
1993. See Manuel Perucho et al., "Ubiquitous Somatic Mutations in
Simple Repeated Sequences Reveal a New Mechanism for Colonic Car-
cinogenesis," *Nature* (363:1993), 558–561. An earlier report in 1992 by
the same lab had previously made reference to the alterations later reported in greater detail in *Nature*. See M. Peinado, "Isolation and Characterization of Allelic Losses and Gains in Colorectal Tumors by
Arbitrarily Primed Polymerase Chain Reaction," Proceedings of the National Academy of Sciences (89:1992), 10065–10069.

    5. Thomas D. Petes et al., "Destabilization of Tracts of Simple Repet-
itive DNA in Yeast by Mutations Affecting DNA Mismatch Repair," *Nature* (365:1993), 274–276.

    6. Thomas Kunkel, "Slippery DNA and Diseases," *Nature* (365:
1993), 207–208.

    7. Richard Fishel et al., "The Human Mutator Gene Homolog MSH2
and Its Association with Hereditary Nonpolyposis Colon Cancer," *Cell*
(75:1993), 1027–1038.

    8. Fred S. Leach et al., "Mutations of a MutS Homolog in Hereditary
Nonpolyposis Colorectal Cancer," *Cell* (75:1993), 1215–1225.

    9. Rick Weiss, "Gene for Colon Cancer Identified; Scientists Foresee
Simple Blood Test for Hereditary Forms of Disease," *The Washington Post*
(December 3, 1993), A1.

12: THE MOTHER OF ALL TUMOR SUPPRESSORS

    1. M. Serrano et al., "A New Regulatory Motif in Cell-cycle Control
Causing Library Specific Inhibition of Cyclin D/CDK4," *Nature*
(366:1993), 704.

    2. Myriad Genetics, Inc., *Private Placement Memorandum*

    3. Lisa Canon-Albright et al., "Assignment of a Locus for Familial
Melanoma. MLM to Chromosome 9p13-p22," *Science* (258:1992), 1148.

    4. M. Serrano.

    5. In an interview Beach acknowledged that Cold Spring Harbor had
filed a patent claim for p16 when it was discovered and that Beach believed Mitotix was interested in licensing the patent rights.

    6. Kamb had uncovered another gene in the same region that also
was defective in many tumor cell lines and which he initially called
MTS2. Coincidentally, MTS1 also served as an abbreviation for the
gene's other appellation, the (M)other of all (t)umor and (s)uppressors.

7. Feinstein says he kept all information about the two companies confidential, and Skolnick says he believes that's true. Beach, however, expressed some bitterness toward Feinstein, saying in an interview that he believed Feinstein "stalked" Beach during this period, monitoring his movements and alerting Myriad as to his activities. Feinstein denies this and there is no evidence it is true.

8. Stillman declined to confirm the conversation, saying only that such talks were private. Beach and others say it occurred. Moreover, within days researchers at Myriad received calls from investigators at several labs asking if Myriad had found the melanoma gene. "Somehow the news got out," says Skolnick. "Was it Stillman? I don't know."

9. By now, news that Myriad had made a major finding related to p16 was bouncing around the AACR, although few details were known for certain.

10. Jean Marx, "New Tumor Suppressor May Rival p53," *Science* (264:1994), 344.

11. *Science*, as with many top scientific journals, such as the *New England Journal of Medicine*, the *Journal of the American Medical Association*, and the British publications *Lancet* and *Nature*, customarily sends journalists advance copies of the next issue a few days to a week ahead of the publication date. The prepublication copies give reporters time to digest the science, set up interviews with the papers' authors, and even call other scientists in the field for comment.

The journals stipulate that reporters abide by an embargo in order to receive the advance copies, thus setting up an informal "gentleman's agreement" of confidentiality between those in the press who get the early editions and the journal editors.

The journal editors say the embargoed release of the research allows for an organized delivery of important news and assures that scientists present their research to the public only after it has been carefully reviewed by their scientific peers prior to publication. The embargoed release time also spurs news coverage of scientific arcania embedded in the ofttimes understated jargon of a research paper. *Science, Nature, Lancet,* and other journals do this by highlighting and summarizing the coming week's contents via headlined news releases often faxed directly to science journalists. In recent years this practice of flagging a journal's noteworthy contents has increased sharply, especially among journals carrying medical science, perhaps a reflection of the public's rising interest in health-related news. At the same time, reporters often are also inundated by press releases from the academic institution that is home to the publishing scientist, or by a corporation whose research or experimental product is the subject of the paper.

And, finally, like many commercial periodicals, the journals compete intensely for subscribers, advertising, and media attention. They help generate interest among these groups by encouraging scientists with significant advances to publish in their pages, and then work hard to generate interest in the research by admonishing the scientists to keep the exact details of their findings from the public until the prepublication copies are distributed to reporters. As a result, many scientists refuse to discuss experiments or studies with journalists who may have gotten wind of the research prior to a journal's release, even though the actual work described in the research paper was completed weeks or months prior to publication. Scientists take this warning seriously. That's because, aside from the public notice, publication in an esteemed journal is the principal currency of the scientific marketplace. It is how scientists are measured by prospective employers, by grant providers, and by their peers. Thus, they are loathe to upset the editors.

This all makes for a high-stakes game in which the scientists and the journals aggressively court the media's attention, but only on highly restrictive terms. Reporters, on the other hand, often consider it a coup to break the journals' stranglehold on scientific news by uncovering and writing about a discovery prior to receiving the embargoed notification. Thus, breaking an embargo, either by a reporter who has somehow gotten hold of research from a source other than the journal, or by the journal itself, is considered a noteworthy development. It happens rarely, but each time carries with it details that reflect the intense competitive pressure on the journals, the scientists, and the media to control and shape the news.

The problem of delicately balancing control and publicity of research is exacerbated when the research in question involves a product, such as a drug or vaccine, being developed by a for-profit company. Should that company be publicly traded, news of the research for an upcoming potentially big-selling new medicine, say, is often considered inside information. An investor armed with such insider knowledge could buy or sell shares well in advance of the rest of the public.

12. In an interview several weeks later, Vogelstein said his quote, calling the discovery one of "phenomenal importance," was correct, but misrepresented his understanding of the nature of Kamb's discovery. Vogelstein hadn't actually seen Kamb's paper. Like many other scientists in the field, he had heard the rumors swirling about that week, and he assumed that Kamb had found the melanoma-predisposing gene and that it was the same as Beach's p16 gene. "It's not Jean's fault," he said in the in-

terview. "She asked me if I was aware of the research and I said, 'yes.'"
He conceded his comment may have given the find a cachet it didn't yet
deserve. Although he later said the gene "certainly is as important as
Kamb originally claimed."

13. In an interview with Curt Harris, Jean Marx was told that the
Harris and Beach labs had found damaged p16 genes in tumor cell lines,
confirming data that convinced her that Kamb's research was legitimate
and impressive. In separate interviews with me on April 12 for an article
I was preparing on the discovery for the *Wall Street Journal's* April 15 issue, Beach and Harris also confirmed the same thing without providing
details. Both Beach and Harris say they received copies of Kamb's study
faxed to them by reporters seeking comment.

14. Tsutomu Nobori et al., "Deletions of the Cyclin Dependent Ki-
nase-4 Inhibitor Gene in Multiple Human Cancers," *Nature* (368:1994),
753–756.

13: THE ROLLER COASTER

1. Beth Newman, "Inheritance of Human Breast Cancer: Evidence
for Autosomal Dominant Transmission in High Risk Families," *Proceedings of the National Academy of Sciences* (85:1988), 3044–3048.

14: RAS

1. Nancy E. Kohl et al., "Selective Inhibition of ras-Dependent
Transformation by a Farnesyltransferase Inhibitor," *Science* (260:1993),
1937; Guy L. James, "Benzodiazepine Peptidomimetics: Potent Inhibitors
of Ras Farenesykation in Animal Cells," *Science* (260:1993), 1937–1941.

2. Nancy E. Kohl et al., "Protein Farnesyltransferase Inhibitors Block
the Growth of ras-Dependent Tumors in Nude Mice," *Proceedings of the
National Academy of Sciences* (91:1994), 9141–9145.

3. Nancy E. Kohl et al., "Protein Farnesyltransferase."

4. *Values & Visions: A Merck Century* (Rahway, N.J.: Merck & Co.,
1991), 97.

5. In 1993, Varmus was named head of the National Institutes of
Health; in 1995 he asked his longtime friend, associate, and cowinner of
the Nobel Prize, Michael Bishop, to argue before Congress the case for
pumping a steady flow of dollars into "basic" biomedical research. Bishop
specifically cited the cancer gene hunt in his argument. In his testimony
he said, "We have a powerful new view of cancer. The seemingly count-
less causes of cancer—cigarette smoke, sunlight, asbestos, chemicals,
viruses, and many others—all these may work in a single way, by playing
on a genetics keyboard, by damaging a few of the genes in our DNA, by

jamming the accelerators and removing the brakes of our cells. An enemy has been found and we are beginning to understand its lines of attack. And it all began with a chicken virus." From testimony before the United States House of Representatives, Labor/HHS/Education Appropriations Subcommittee, Feb. 28, 1995.

6. An informative and lively account of this inquiry is carried in Natalie Angier's book, A Natural Obsession (New York: Houghton Mifflin Co., 1988), 96–117.

7. Barry Werth, The Billion Dollar Molecule (New York: Simon & Schuster, 1994), 130.

8. Joseph L. Goldstein and Michael S. Brown, "Regulation of the Mevalonate Pathway," Nature (343:1990), 425–430.

9. Ibid., 430.

10. Y. Reiss et al., "Inhibition of Purified p21ras Farnesyl:Protein Transferase by Cys-AAX Tetrapeptides," Cell (62:1990), 87.

11. Marilyn M. Chase, "Genentech, Merck Report Cancer Advance," The Wall Street Journal (June 25, 1993).

15: FIRST, DO NO HARM

1. Rachel Stein asked that her real name not be used to protect her family from possible loss of health or life insurance. In addition, she said, "I'm not certain my other family members would like to see themselves put at any greater risk than they already are. I do believe my story may help others. I wish I could be more [open], but I can't take the chance."

2. The National Institutes of Health officially designated the National Center for Human Genome Research as a full-fledged "institute" in late 1996, giving it wider responsibilities, increased resources and influence accorded the other institutes in the NIH family.

3. S. Lehrman, ". . . As Concern Grows over Screening," Nature (384:1996), 297.

4. O. I. Olopade, "Genetics in Clinical Cancer Care—The Future Is Now," New England Journal of Medicine (335:1996), 1456.

5. J. P. Struewing et al., "The Carrier Frequency of BRCA1 185delAGds Approximately One Percent in Ashkenazi Jewish Individuals," Nature Genetics (11:1995), 198–200.

EPILOGUE

1. Lisa Canon et al., "Genetic Epidemiology of Prostate Cancer in the Utah Mormon Genealogy," *Cancer Survey* (1:1982), 47.

2. J. R. Smith et al., "Major Susceptibility Locus for Prostate Cancer on Chromosome 1 Suggested by a Genome-Wide Search," *Science* (274:1996), 1371–1374.

3. Thomas Maugh, "Hunt for Prostate Cancer Gene Advances," *Los Angeles Times* (November 24, 1996), A42.

Bibliography

The material in this book is largely the result of reporting and research I conducted as part of my job as a medical and science reporter for *The Wall Street Journal*.

What follows is a list of sources I drew upon in some manner over the several years I spent researching and writing this book. Most of the books listed here provided me with background material or guidance. They are cited for the reader who is curious about the published sources for scientific and historical facts that are not the result of my own research. In addition, they are listed for those interested in pursuing various subjects touched upon in my text.

Some of the following sources are textbooks, relied upon largely for scientific accuracy and as educational guides. Some are nonfiction narratives, relied upon for their independent reporting unavailable to me, or to double-check my own research. Whenever any of the sources provided a specific quote or an exceptionally lengthy or proprietary reference that I was unable to acquire though my own research, it was cited in the "Notes" section.

Angier, Natalie. *Natural Obsessions: The Search for the Oncogene*. Boston: Houghton Mifflin Co., 1988.

Bishop, Jerry E., and Michael Waldholz. *Genome: The Story of Our Astonishing Attempt to Map All the Genes in the Human Body*. New York: Simon & Schuster, 1990.

Cavenee, Webster, ed. *Recessive Oncogenes and Tumor Suppression, Current Communications in Molecular Biology*. Cold Spring Harbor, N.Y.: Cold Spring Harbor Laboratory Press, 1989.

Cold Spring Harbor Laboratory Symposium on Quantitative Biology. *Molecular Biology of Cancer: June 1–8, 1994*. Cold Spring Harbor, N.Y.: Cold Spring Harbor Laboratory Press, 1994.

Cooke-Deegan, Robert. *The Gene Wars: Science, Politics and the Human Genome*. New York: W. W. Norton & Co., 1994.

Cowell, J. K., ed. *Molecular Genetics of Cancer*. Oxford: Bios Scientific Publishers, 1995.

Davies, Kevin, and Michael White. *Breakthrough: The Race to Find the Breast Cancer Gene*. New York: John Wiley & Sons, 1996.

Griffiths, Anthony J., et al. *An Introduction to Genetic Analysis*, Fifth Edition. New York: W. H. Freeman & Co., 1993.

Micklos, David A., and Greg A. Freyer. *DNA Science: A First Course in Recombinant DNA Technology*. Cold Spring Harbor, N.Y.: Cold Spring Harbor Laboratory Press, 1990.

Murray, Andrew, and Tim Hunt. *The Cell Cycle*. New York: W. H. Freeman & Co., 1993.

Neel, Benjamin G., and Ramesh Kumar. *The Molecular Basis of Human Cancer*. Mt. Kisco, N.Y.: Futura Publishing Co., 1993.

Pines, Maya, ed. *Blazing a Genetic Trail*. Bethesda, Md.: Howard Hughes Medical Institute, 1991.

Rothwell, Norman V. *Understanding Genetics: A Molecular Approach*. New York: Wiley Liss & Co., 1993.

Varmus, Harold, and Robert A. Weinberg. *Genes and the Biology of Cancer*. New York: Scientific American Library, 1993.

Watson, James D. *The Double Helix: A Personal Account of the Discovery of the Structure of DNA*. A Norton Critical Edition, ed., Gunther S. Stent. New York: W. W. Norton & Co., 1980.

Weatherall, D. J. *The New Genetics and Clinical Practice*. Oxford: Oxford University Press, 1991.

Weinberg, Robert A. *Racing to the Beginning of the Road: The Search for the Origin of Cancer*. New York: Harmony Books, 1996.

Werth, Barry. *The Billion Dollar Molecule: One Company's Quest for the Perfect Drug*. New York: Simon & Schuster, 1994.

Index

Page numbers in italics refer to diagrams.

Abeloff, Martin, 143
adenomas, *see* polyps
AIDS, 181, 217, 237
Allende, Salvador, 99
American Association of Cancer
    Research (AACR):
    annual meeting of, 201-9
    Beach's press briefing at, 203-8
American Society for Human Ge-
    netics, 162, 201-2
    King's breast cancer findings
    presented at, 42-47
American Urological Association,
    283
amino acids, 23-24
Anderson, David, 106
Angier, Natalie, 151
Antmann, Karen, 271-72
Ashkenazi Jews:
    BRCA1 mutations shared by,
    264-68
    in NCI study, 264-65

Baker, Ben, 138-39
Baker, Sue, 144
balanced translocation, 224

Baltimore Orioles, 171-72
base pair sequences, 23
    mismatched, 167-68
    in mutated vs. normal p53
    gene, 25
    in protein assembly, 23-24
    in ras, 240
    in repeat marker of HNPCC,
    162, 169-70
    restriction enzymes in splicing
    of, 127-28
    see also DNA
Bazell, Robert, 231
Beach, David, 209
    in AACR press briefing, 203-8
    collaboration of Harris and,
    197-98, 205, 207
    in discovery of p16 gene, 187-88
    Kamb's competition with, 192-
    93, 194-208
Berman, Edgar, 18-19
Bethesda Naval Hospital, 17
Biesecker, Barbara Bowles, 78-85
biotechnology companies:
    academic researchers in, 227
    see also individual companies

Bishop, J. Michael, 11, 237, 239-240

Black, Donnie, 212

bladder cancer:
  of Humphrey, 15-27, 68-69, 278
  p53 gene in, 25-27, 146, 278
  test kit for, 278

Bodmer, Walter, 56

Botstein, David, 126-30

Bowcock, Anne, 43

Brady Urological Institute, 280

brain cancer, 235
  deaths from, 277
  PTEN gene in, 277-78

Brandeis University, 95

BRCA1 gene, 211-32, 255-75, 278, 285
  coding region of, 230, 262
  identification of, 230-31
  mutations of, in Ashkenazi Jews, 264-68
  patenting of, 231, 262
  see also chromosome 17

BRCA2, 258, 261, 265

breast cancer, 14, 20, 101, 137, 235, 280
  as "acquired" disease, 104
  cases of, 34
  deaths from, 34
  early-onset type of, 45-46, 218-219
  familial, 29-30, 45, 46-47, 53-54, 101-2, 104, 105-7, 119, 214-15, 216-19, 222-23, 223, 228-29, 230, 231, 258, 261, 267-68, 269-72
  in Family 15, 33-41, 57-87
  gene markers in, 40, 73-75, 215, 218-19, 222-23, 223
  health insurance and, 93, 257
  male, 119
  mutated p53 gene in, 146
  nonfamilial, 258-59
  prophylactic surgery for, 33, 34, 35, 68, 82, 256, 259-60, 268, 270-73, 274
  see also BRCA1 gene; King, Mary-Claire; Lynch, Henry; human nonpolyposis colon cancer

breast self-exams, 259

Broca, Paul, 105

Brown, Michael S., 249-50, 251

Burt, Randall, 121-22

California, University of (Berkeley), 42, 43, 95, 104-9, 110, 114, 208

California, University of (Irvine), 168

California, University of (San Diego), 210, 257

California, University of (San Francisco), 100-104, 105, 237

Calzone, Kathy, 59, 60, 61, 68-69, 73, 76, 77, 78-79, 80, 81
  in counseling of Family 15, 82-87

Cambridge University, 90

cancer:
  deaths from, 13, 28, 64
  epidemiologists and, 101, 105-7
  gene-based therapies for, 30-31
  genetic testing for, 27, 30-31, 35, 151-52, 158, 191, 255-75
  multistep model of, 136-47
  mutated p53 gene in, 19-20, 25-27, 28, 143-47, 149-52
  ras gene in, 235, 236, 240-41
  survival rates in, 27

theories of, 136
tumor suppressor theory of, 132-41
see also Family 15; oncogenes; specific types of cancer
Cannon-Albright, Lisa, 227, 280
Caplan, Arthur, 272-73
carcinoma in situ, 18
carcinomas, 139
Carleton College, 95
Cartwright, George, 123-24
Cavalli-Sforza, Luca, 44
Skolnick and, 115-17, 118, 119
CDK4 (cyclin-dependent kinase 4) (enzyme), 187-88
Cell, 176, 179, 196
cell cycles, 192-93
cell lines, 188-89, 193, 198
immortalized, 189
chemotherapy, 30, 64, 236, 250, 271
chimpanzees, genetics of humans vs., 98-99
chromosome 1, 285
chromosome 2, 162, 169, 170, 171, 172, 175, 182
chromosome 3, 180, 182
chromosome 5, 141, 143, 147, 172
chromosome 6, 126
chromosome 7, 192, 197, 202
chromosome 11, 141, 197
chromosome 13, 137-38
chromosome 17, 47
breast cancer and, 42-56, 73-75, 211-13, 219, 222
colon cancer and, 140, 141, 143
p53 located on, 144
chromosome 18, 140, 141, 143, 147, 150
chromosomes, 74-75

anomalies in, 102-3
translocated, see balanced translocation
see also specific chromosomes
cloning of genes, 38-39, 220-21, 221
coding region in genes, 230, 262
Coffey, Donald, 136, 279-80
Cold Spring Harbor Laboratory, 48, 51, 186-87, 192-93
symposium at, 198-99, 200
Cole, Richard, 209
Collins, Francis, 35-36, 60, 73, 77, 81, 82, 111, 150, 151, 178, 190, 198, 226, 229, 259
collaboration of King and, 47-52, 89, 212, 213, 219-25, 229-31
in counseling of Family 15, 82-87
cystic fibrosis gene mapped by, 38-39
on genetic testing, 257-58
HPC-1 gene mapped by, 283-86
Weber and, 38-39, 61, 229
colon cancer, 14, 53, 101, 120, 121-22, 235, 280
deaths from, 139
familial, 147, 148-64, 171, 173, 175-76
gene markers in, 140-42, 162, 169-70, 172-74
mutated p53 gene in, 19-20, 143-47, 149-52
ovarian and uterine cancer linked to, 151-52, 177
pedigrees in de la Chapelle's study of, 152-55
polyp growth in, 139
Vogelstein's research on, 19-20, 138-83

Columbia-Presbyterian Medical
    Center, 271
Columbia University School of
    Medicine, 243, 277
Colwell, Robert, 99-100, 104,
    108-9
Congress, U.S., research funding
    of, 42, 89, 90, 239-40
Conneally, P. Michael, 119
cortisone, 242
Creighton University School of
    Medicine, 53
Crichton, Michael, 25
Crick, Francis, 23, 47, 52
cystic fibroid disease, 65
cystic fibrosis, 47, 102, 104, 111,
    129, 215, 217, 244
    Collins's mapping of gene for,
        38-39
cystoscopy, 17

Dana Farber Cancer Institute,
    145, 150, 165
Davis, Ronald, 126-30
de la Chapelle, Albert:
    collaboration of Vogelstein
        and, 153-65
    Finnish pedigrees and, 152-55
Denmark, national cancer study
    in, 105, 107
diabetes, 24, 43, 103, 168
DNA (deoxyribonucleic acid), 13
    bases of, see base pair sequences
    as component of genes, 23-24
    enzymes in manipulation of,
        126-27
    in "fingerprinting" technology,
        25, 49, 111
    laboratory amplification of, 25
    "quality control" proteins in,
        149-51
    recombination of, 168, 172-73,
        221-22, 223, 224, 228-29
    splicing of, 127-28
    structure of, 23-24
DNA repair genes, see mismatch
    repair genes
Double Helix, The (Watson), 48
Down's syndrome, 103
Duke University, 173
dysplastic nevi, 191

Edwards, Corwin, 123-24
electrophoresis gels, 230-31
Eli Lilly & Co., 91, 226
endocrine neoplasia, 53
epidemiologists, epidemiology,
    100-102
    cancer and, 101, 105-7
    and risk factors of heart disease,
        100-101
Erozan, Yener S., 19, 21
estrogen, EDH17B gene and, 225
"Eve," 96
"Evolution at Two Levels in Hu-
    mans and Chimpanzees"
    (King and Wilson), 98
eye cancer, childhood, see
    retinoblastoma

familial adenomatous polyposis
    (FAP), 150, 152
    gene in, 147, 155-56
families:
    breast cancer in, 29-30, 33-41,
        45, 46-47, 53-54, 57-67,
        101-2, 104, 105-7, 119, 214-
        15, 216-19, 222-23, 223,
        228-29, 230, 231, 258, 261,
        267-68, 269-72
    colon cancer in, 147, 148-64,
        171, 173, 175-76

prostate cancer in, 280, 281-85
Family One, 216-17
  pedigree of, 217
Family 7, pedigree of, 222, 223
Family 15 (Stone family), 33-41,
    57-87, 260-61
  gene markers in, 40, 73-75
  genetic counseling and, 59, 76-
    87
  ovarian cancer in, 63-64, 81
  pedigree of, 86
  prostate cancer gene in, 79-80
Family 2082, 228-29, 230, 231
family trees, see pedigrees
farnesyl tranferase (FT), 249-52
  see also ras gene
Fearon, Eric, 140, 143, 147
Feinstein & Partners, 194-95
Finland, colon cancer study in,
    152-55
Fishel, Richard, 166-83
Food and Drug Administration
    (FDA), 269
Fortune, 244
Fountain, Jane, 202
Friedman, Lori, 44, 93, 229
Friend, Stephen, 92, 145, 163
Frost, John D. (Jack), 17-19
Futreal, Andy, 229, 230

GAP, 247-48
Garber, Judy, 173-76
Genbank, 184-86, 188, 193
Genealogical Society of Utah,
    116, 119
gene-based therapies, 30-31
gene hunters, 22-23
  collaboration of, 39, 47-52, 89,
    148-49, 153-65, 172, 173,
    174, 176-77, 212, 213, 215-
    16, 218, 219-25, 229-31

lab techniques of, 39, 172,
    188-89, 215-16, 218, 220-
    21, 230
  in private industry, 91, 181-83,
    226-27
  research subjects and, 39-40
  see also research, researchers;
    molecular biologists, molec-
    ular biology
gene hunters, competitiveness of,
    44, 50-51, 52, 56, 89-93,
    134, 143, 171
  in BRCA1 search, 211-13, 224-
    32
  in HPC-1 search, 279, 285-86
  in MSH1 search, 176-80
  in p16 search, 186, 188, 189-
    90, 194-209
  in search for ras blocking drugs,
    236-37, 250-54
  in search for ras-GAP in-
    hibitors, 248, 249-50
gene markers, 112, 113-14
  in breast cancer, 40, 73-75, 215,
    218-19, 222-23, 223
  in colon cancer, 140-42, 162,
    169-70, 172-73
  in hemochromatosis, 111-12,
    122-26
  in hemophilia, 108
  in Huntington's disease, 76-77,
    111, 129
  mapping of, see genes, mapping
    of
  polymorphic, see restriction-
    fragment-length polymor-
    phism markers
  in prostate cancer, 284-85
  and recombination of DNA,
    172-73, 222, 224, 228-29
  repeats as, 162, 169, 173, 174

gene markers *(cont.)*
  in test for mutated gene's pres-
    ence, 74-75
Genentech, 250, 251, 252
genes:
  cancer-causing, *see* oncogenes
  cloning of, 38-39, 220-21, *221*
  coding region in, 230, 262
  identification of, *see* genes,
    cloning of
  linkage of, *see* gene markers
  mapping of, 38-39, 42-56, 60,
    103, 107, 126, 148-64, 165-
    70, 190, 202, 211, 213-19,
    283-86
  patenting of, 181, 188, 194,
    231, 262
  structure of, 23-24
  see also gene markers; *specific
    chromosomes, genes and types
    of genes*
genes, mutated, 22
  diseases linked to, 24, 43, 47
  markers in test for presence of,
    74-75
  see also oncogenes; p53 gene,
    mutated
gene splicing, 127-28
genetic counseling, counselors, 78-
  85, 270-72
geneticists:
  and "inherited" vs. "acquired"
    diseases, 102-4, 112
  see also gene hunters; molecu-
    lar biologists, molecular bi-
    ology
Genetics and IVF, 267-68
genetic testing, genetic tests, 27,
  30-31, 35, 158
  for breast cancer, 255-75
  counseling and, 270-72

ethics of, 31, 151-52, 256-60,
  268, 272-73, 274
Family 15 and, 76-87
full-gene sequencing in, 262-64
in hemochromatosis, 125-26
for Huntington's disease, 76-77
interpretation of risk factors in,
  269-70
Jewish women and, 266-67
marketing of, 257-58, 261-63,
  266-67
Myriad Genetics Inc. and, 257,
  261-64, 268-69
patient empowerment and,
  258, 274-75
standards for, 269
genome, 42, 103
  see also National Institute for
    Human Genome Research
germ cells:
  balanced translocation in, 224
  recombination in, 168, 221-22
Gibbs, Jackson (Jay), 245-53
  in cloning of GAP gene, 248
Gilbert, Walter, 226
Glamour, 88, 93
Glaxo, 244
Goldberg, Barbara, 202-3
Goldgar, David, 227-31
Goldstein, Joseph L., 249-50, 251
Good Morning, America, 179
Grandmothers of Plaza de Mayo,
  49, 217
Green, Jane, 148-62
  collaboration of Vogelstein
    and, 148-49, 156-63, 173
  Ishmael Mast's family history
    collected by, 148-49, 158-
    62
  and Von Hippel-Lindau dis-
    ease, 157-58

Hall, Jeff, 215-16, 217-20
Hamilton, Stanley, 138, 139-41, 173
Harris, Curt, 145, 182-83, 197-98
    in collaboration with Beach,
        197-98, 205, 207
Harvard University, 92, 145, 149,
        150, 163, 166, 168, 181,
        226, 238
    School of Medicine of, 246
Haseltine, William, 181-83
Health and Human Services De-
        partment, U.S., 178
heart disease, 24, 43, 120-21, 244
    risk factors for, 100-101
Hedrick, Laura, 156-57
Helsinki, University of, 152
hemochromatosis, 122-26
    gene markers in, 124-26
hemophilia, 108
Hirschfeld-Bartek, Judi, 266-67
hMLH1 gene, 180-83
hMSH1 gene, 176-80
Holtzman, Neil, 152, 268-69, 274
Hopkins Bowel Tumor Working
        Group, 138-42
Howard Hughes Medical Institu-
        tions, 39
Hruban, Ralph H., 18-22
    analysis of Humphrey's cancer
        cells by, 21-22, 25-27
Human Genome Project, see Na-
        tional Institute for Human
        Genome Research
Human Genome Sciences, Inc.
        (HGS), 181-83
human leukocyte antigens (HLA),
        124-26
human nonpolyposis colon cancer
        (HNPCC), 53, 148-64
    cloning of gene in, 170-83
    in Finnish families, 152-55

gene markers in, 162, 169-70,
        172-74
    in Green's Newfoundland doc-
        umentation, 148-49, 156-62
    mapping of gene in, 148-64,
        165-70
human prostate cancer gene
        (HPC-1), mapping of, 283-
        86
Humphrey, Hubert H., 15-27, 68-
        69, 278
    biopsy cell slides of, 17-19, 20-
        22, 25-27
    death of, 18
    as presidential candidate, 16-17
Humphrey, Muriel, 21-22
Huntington's disease, 215
    gene markers in, 76-77, 111,
        129
    presymptomatic testing and, 77
hypertension, 104, 244

immune system, 136
Indiana, University of, 119, 244
"inherited diseases," geneticists
        on, 102-4, 112
The Institute for Genetic Re-
        search (TIGR), 181
insulin, 24, 168
International Agency for Research
        on Cancer, 52
International Breast Cancer Link-
        age Consortium, 90
International Cancer Research
        Fund, 55-56, 92, 143
Isaacs, Sally, 283-84
Isaacs, William "Billy," 283-84

Jackson Laboratory, 155
Jews, Ashkenazi, see Ashkenazi
        Jews

Johns Hopkins, 15-28, 131, 135,
    152, 155, 279
    Brady Urological Institute of,
        280
    Oncology Center of, 143
    School of Medicine of, 136,
        268
Johnson, Lyndon Baines, 6
Jurassic Park (Crichton), 25

Kamb, Alexander (Sasha), 184-
    210, 211, 227
    in AACR press briefing, 206-8
    Beach and, in p16 gene re-
        search, 192-93, 194-208
    see also p16 gene
Kabak, Michael, 257, 260, 270
King, Mary-Claire, 87, 100-109,
    198, 237, 259, 283, 286
    Collins's collaboration with,
        47-52, 89, 212, 213, 219-25,
        229-31
    doctoral dissertation of, 97-99
    existence of breast cancer gene
        proved by, 42-56, 60, 73,
        211, 213-19
    lab of, 42, 43, 45-46, 49, 51, 89,
        213, 215, 217-18, 229-30
    Lenoir and, 55-56, 89
    as population geneticist, 50
    on prophylactic surgery, 82, 270
    on race for breast cancer gene,
        93
    in search for breast cancer
        gene, 88-94, 104-9, 213-32
    as seen by colleagues, 93-94,
        218
    Watson and, 48-50
    Weber and, 89
    as Wilson's protege, 95-99, 102
Kinzler, Kenneth, 131, 133, 141,
    146, 150, 155, 171, 177,
        180, 181, 182, 183
Klausner, Richard, 278, 284-85
Knudsen, Alfred, 137-38, 144
Kolodner, Richard, 149, 150, 165-
    83
    and discovery of MSH, 165-68
    and Vogelstein in cloning of
        HNPCC gene, 170-83, 212
Koshland, Daniel, 195-96, 208
Kravitz, Kerry, 125
Kunkel, Tom, 174

Lander, Eric, 151
Lane, David, 143-44, 146
Lark, Gordon, 117-18
Lederberg, Joshua, 114, 115
Legionnaires' disease, 100
Lenoir, Gilbert, 52-56
    King's findings corroborated by,
        55-56, 89
    Lynch and, 53-54, 107
leukemia, 101
Levine, Arnold, 143-44
Levy, Dan, 156-57
lip cancer, 119
Liskay, Michael, 179-81, 183
Los Angeles Times, 285
lumpectomy, 72
lung cancer, 20, 29, 34, 101, 103,
    235, 279
Lynch, Eric, 229
Lynch, Henry, 53-54, 82, 107, 108,
    112, 113, 162-63, 172, 173,
    214
Lynch, Patrick, 163, 172
Lynch syndrome, see human non-
    polyposis colon cancer
Lyons, University of, 52

McCormick, Frank, 147, 247-48

and human GAP cloning, 248
McKusick, Victor, 155
mammogram, 62-63
Marsters, James C., Jr., 252
Martin, Virginia, 58-59
Marx, Jean, 204-6
Massachusetts, University of, 129
Massachusetts General Hospital,
    145, 163, 238
Massachusetts Institute of Technol-
    ogy (MIT), 126, 151, 202, 245
mastectomy:
    bilateral, 72, 259
    radical, 72
Mayo Clinic, 34, 69, 151, 169
M. D. Anderson Cancer Center,
    46, 64-65, 106, 137, 163
Mecklin, Jukka-Pekka, 152-54
media, gene discoveries and, 88,
    151, 178-79, 202-9, 216,
    231, 252, 285
medical ethicists, 272
    see also genetic testing, genetic
    tests, ethics of
melanoma, 120
    deaths from, 190-91
    p16 gene and, 184-210
Meldrum, Peter, 226
Memorial Sloan-Kettering Cancer
    Center, 18, 22, 27, 193, 248,
    258, 264
Mendel, Gregor, 105, 106
Merck, 233-54
    academic environment of, 241-
    42
    cancer research at, 233-37,
    245-55
    drugs developed by, 233-34,
    242
    enzyme-targeting drug strategy
    of, 243-44, 249

and prototype ras-blocking
    drug, 235-37, 250-54
Merck, Friedrich Jacob, 241
Merck, George, 241
Merck, George W., 241-42
Michigan, University of, Medical
    Center of, 33-41, 57-61, 68-
    69, 72, 73-74, 77-87
Miki, Yoshio, 230
Milken, Michael, 279, 285
Minna, John, 237
mismatch repair genes, 232
    in bacteria, 149, 173-74, 179-
    82
    defective, 150-51
    in humans, 149-51, 168, 174-
    83
    in yeast, 149, 167-70, 175, 179
    see also p16 gene; p53 gene
Mitotix, Inc., 194-95
Modrich, Paul, 173-74, 176, 179
molecular biologists, molecular bi-
    ology, 22, 39-40, 50, 78, 89-
    93, 135, 192, 234
    see also gene hunters
Molino, Patricia, 202-8
Mormons:
    family size of, 113, 120, 121
    genealogy records of, see Utah
    genealogy
    male breast cancer in, 119
MSH gene, 165-68
    DNA repair mechanism in,
    167-68
MSH1, 168, 177-80
MTS1, see p16 gene
muscular dystrophy, 47, 102, 104,
    111, 129, 215
mutL gene, 179-82
Myriad Genetics Inc., 91-92
    BRCA1 discovered by, 230-31

Myriad Genetics Inc (*cont.*)
  BRCA1 research at, 211-13, 226-
    32
  BRCA1 test offered by, 257,
    261-63
  formation of, 91, 226
  p16 gene patented by, 188, 194
  p16 gene research of, 184-210
  Skolnick's stake in, 189-90, 191

Nader, Ralph, 44, 97
Narod, Stephen, 264
National Academy of Sciences,
    243
National Cancer Institute, 135,
    145, 182, 197, 204, 207,
    264, 278, 284
  breast cancer study of (1982),
    214
  breast cancer study of (1997),
    259
  Scolnick at, 239-41
National Institute for Human
    Genome Research, 22-23,
    42, 88, 111, 129, 229, 257-
    58, 284-85
National Institute of Environmen-
    tal Health Sciences, 229
National Institutes of Health, 39,
    89, 172, 179, 180, 197, 212,
    226, 231, 234, 243, 245
  National Cancer Institute of,
    see National Cancer Insti-
    tute
  Skolnick's grant and, 118-20
Nature, 196, 210, 240
  p16 gene in, 187, 193, 197
  ras-inhibiting enzyme in, 250
Navia, Manuel, 244
neurofibromatosis, 47, 224
New England Journal of Medicine,

    26, 141-42, 260
Newfoundland, genetic disease
    clusters in, 157-62
Newman, Beth, 215, 218
"New Tumor Suppressor May Ri-
    val p53" (Marx), 203-6
Newsweek, 99
New York Times, 88, 151, 209
New York University, 253
Nirenberg, Marshall, 238
Nixon, Richard M., 17
Nobel Prize, 90, 131, 239, 242, 249
Nowell, Peter, 141-42

Offitt, Kenneth, 258, 264-66, 274
Oliff, Allen, 248-53
oncogenes, 14, 24, 102, 239-40
  in bird virus, 239-40
  in bladder cancer, 15-27
  in breast cancer, 35-41, 42-56,
    68-87, 88-94
  in colon cancer, 19-20, 24, 141
  mutant p53 and, 16, 19, 28,
    144-46
  mutation of, 137, 139
  screening for, *see* genetic test-
    ing, genetic tests
  single point mutation and, 240
oncologists, 72
OncorMed, 257, 269
Onyx Pharmaceuticals, 147
oophorectomy, 151, 259, 268, 273
Orange County Register, 209
osteoporosis, 244
Ostermeyer, Beth, 229
ovarian cancer, 54, 55, 63-64, 81,
    151-52, 258, 274-75
  see also oophorectomy

Paisely, David, 244-45
pancreatic cancer, 235

Papadopolous, Nick, 134, 160, 174-75, 182
Pap smear, 63
Parade, 284
Parma, University of, 115-16
Parsons, Ramon, 164, 180, 182
    PTEN gene discovered by, 277, 278, 279
patenting of genes, 181, 188, 194, 231, 262
Patent Office, U.S., 194
Pauling, Linus, 209
pedigrees, 107, 218
    of Family One, *217*
    of Family 7, *222, 223*
    of Family 15, 86
    of Finnish colon cancer families, 152-55
    Parma, 115, 117
    of Walsh's prostate cancer patients, 282-84
Pennsylvania, University of, 135, 229, 243, 273
    Center for Bioethics of, 272
    Medical Center of, 259, 264
Perucho, Manuel, 174
Petersen, Gloria, 173
Petes, Tom, 174, 180
Petrakis, Nicholas, 105-6, 108, 120
pharmaceutical companies, ras-blocking drugs and, 236, 237, 253-54
p16 gene, 184-210
    CDK4 inhibitor and, 187-88
    and media coverage, 203-9
    patenting of, 188, 194
    tumor cell lines and, 188-89, 193, 198
p21 gene, 176, 196-97
p53 gene, 131-47
    as anticancer medicine, 146-47

in cell-dividing process, 145-46
location of, on chromosome, 17, 114
as repair gene, 145-46, 150-51
as tumor suppressor, 132, 145-56, 197-98, 278
p53 gene, mutated, 19, 28, 278, 279
    in bladder cancer, 15-16, 25-27, 146, 278
    in breast cancer, 146
    in colon cancer, 19-20, 143-47, 149-52
Phil Donahue Show, 284
polymerase chain reaction (PCR), 216
polymorphic gene markers, 107-8, 127-30
    restriction enzymes in, 127-28
    see also restriction-fragment-length polymorphism markers
polyps, 139, 140
    see also familial adenomatous polyposis
Ponder, Bruce, 90, 91-92
population genetics, 50
    see also King, Mary-Claire; Skolnick, Mark; Utah genealogy
positional cloning of genes, 39, 220-21, *221*
    YACs in, 220
predictive testing, *see* genetic testing, genetic tests
presymptomatic testing, *see* genetic testing, genetic tests
Princeton University, 143
private industry, gene hunters in, 91, 181-83, 226-27
    see also individual companies

probands, 61, 105, 106
Proceed with Caution (Holtzman), 269
prostate, enlargement of, 280-81
prostate cancer, 62, 79, 279-86
    deaths from, 279
    diagnostic test for, 281
    familial, 280, 281-85
    gene markers in, 284-85
    surgery for, 280-81
proteins, base pair sequences and, 23-24
PSA, 281
PTEN gene, 277-78
Public Health Service, U.S., 238

radiation, 137
radiation therapy, 30, 64, 236, 250, 271
Ragovin, Cathie, 266
ras-blocking drugs, 236-37, 250-54
    in mouse experiments, 252, 253
    side effects of, 252-53
ras gene, 233-54
    FT and, 249-52
    GAP and, 247-48
    mutated, 235, 246
    single point mutation in, 240-41, 247
Reagan, Nancy, 216
recombination of DNA, 168, 221
    in Family 7, 222, 223
    gene markers and, 172-73, 222, 224, 228-29
repair genes, *see* mismatch repair genes
research, researchers, 89-91
    funding of, 90-91, 166, 226
    joint affiliations of, 227
    see also gene hunters

restriction enzymes, 126-27
restriction-fragment-length poly-morphism (RFLP) markers, 128-30, 215, 217
    in colon cancer, 140-42
    in retinoblastoma, 138
retinoblastoma, 144
    RFLP marker in, 138
    "two-hit" theory and, 137-38
Rockefeller University, 242
Russo, Donna, 271

Sanders, Charles, 244
San Francisco Chronicle, 209
sarcomas, 239
Sarrett, Lewis, 242
Schlessinger, Joseph, 253
Science, 54, 98, 151, 183, 277, 285
    FT inhibitors in, 251-52
    HNPCC gene mapping in, 163, 164, 165, 169, 170, 171, 174
    Kamb & Skolnick's p16 paper in, 195-96, 198, 199-200, 201, 206
    p16 gene article in, 203-6, 207-210
Science Watch, 131
Scolnick, Edward M., 233-54
    as director of Merck research, 235-55
    at National Cancer Institute, 238-41
    ras gene identified by, 239-40
Shalala, Donna, 178
Shattuck-Eidens, Donna, 196-97, 199-200
    BRCA1 research of, 211-13, 225-32
Sidransky, David, 19-22, 24-27, 155

Sigal, Irving, 246-48
  death of, 248
Simpson, O. J., 25
single point mutation, 240-41, 247
skin cancers, 137
  see also melanoma
Skolnick, Angela, 115, 119, 187,
  191
Skolnick, Joshua, 191
Skolnick, Mark, 89, 110-30, 187-
  210, 211, 237, 259, 280,
  285
  Cavalli-Sforza and, 115-17,
    118, 119
  "fishnet" program of, 117, 124
  as founder of Myriad Genetics,
    Inc., 91, 226
  on genetic testing, 258
  hemochromatosis gene marker
    discovery of, 111-12, 122-26
  Myriad Genetics and, 189-90,
    191
  NIH grant proposal of, 118-20
  parents of, 114
  Utah cancer registry and, 117-
    118, 119-20
  and Utah genealogy, 113, 116,
    119, 120, 122, 124, 190,
    192, 214, 225, 228
SmithKline Beecham, 181
Solomon, Ellen, 55-56, 89, 92-93
Squibb, 244
Stanford University, 114, 115, 126
Star Trek, 229-30
Stillman, Bruce, 198-99, 200
St. John's University, Medical
  Center of, 156, 157-59
Stopfer, Jill, 259, 273
Strong, Louise, 46
Sugen, Inc., 253

Tanh Tran, 230
Tay-Sachs disease, 265
Texas, University of, Southwest-
  ern Medical Center of, 43,
  237, 249
Thibodeau, Stephen, 151, 169-70
Time, 209, 242
Tischler, Max, 242
Todaro, George, 239
Trent, Jeffrey, 172, 284
tumor suppressors:
  p16 gene as, 184-210
  p53 gene as, 145-46, 197-98, 278
  PTEN gene as, 277

USA Today, 209
Utah, University of, 90, 91, 112,
  117, 120, 129, 184, 192,
  227, 280, 285
Utah cancer registry, 117-18, 119-
  20, 225, 228, 280
Utah genealogy, 113, 116-17, 118-
  19, 120, 122, 124, 190, 192,
  214, 225, 228
Utah Population Database, 228
uterine cancer, 151

Vagelos, P. Roy, 241-45, 253
  as head of Merck's research,
    242-47
  as Merck CEO, 247
van der Riet, Peter, 25-26
Varmus, Harold, 239-40
Venter, Craig, 181
Vermont, University of, 166
viruses, 136
  bird, oncogenes in, 239-40
Vogelstein, Bert, 131-47, 190, 196,
  198, 204, 209, 226, 237,
  277, 284, 285-86
  career path of, 135

Vogelstein, Bert (*cont.*)
  colon cancer research of, 19-20,
    138-83
  competitiveness of, 134, 171
  de la Chapelle's collaboration
    with, 153-65
  Green's collaboration with,
    148-49, 156-63, 173
  HNPCC gene mapped by, 148-
    164, 165-70
  and Kolodner, in cloning of
    HNPCC gene, 170-83, 212
  lab of, 19-22, 25-26, 131-34,
    140-41, 144, 147, 160, 164,
    171, 177, 180, 182-83
  multistep cancer model of, 136-
    147
  personal style of, 132-34
  as regarded by colleagues, 131-
    132
  religious background of, 135
Vogelstein, Ilene Cardin, 135-36
Von Hippel-Lindau disease, 157-
  158

Waksman, Selman A., 242
Waldman, Todd, 132
Wall Street Journal, 13, 88, 209,
  231, 252
Walsh, Patrick, 280-86
  and nerve-sparing prostate
    surgery, 281-82
  prostate cancer pedigrees of,
    282-84
Warren, Homer, 118
Washington, University of, 285,
  286
Washington Post, 179, 209

Washington University School of
  Medicine, 243
Watson, James, 23, 47, 48-50, 52,
  198, 229
  as broker of King and Collins's
    collaboration, 48, 50-52
  as Human Genome Project di-
    rector, 51, 88
  King and, 48-50
Weber, Barbara, 35-41, 72-73, 82,
  85-86, 229, 273
  Collins and, 38-39, 61, 229
  and ethical dilemma of genetic
    testing, 57-60, 75-79
  King and, 89
Wei Ding, 230
Weinberg, Robert, 240-41
Weiss, Rick, 179
West Point, Pa., 233
White, Ray, 92, 93-94, 129, 147,
  152, 155
  and human RFLP, 129, 138, 216
Whitehead Institute for Biomed-
  ical Research, 240
Whitmore, Willett, 27
Wilms's tumor, 95
Wilson, Allan, 95-99, 102, 104, 107
  gene analysis in human origin
    theories of, 95-96, 98-99
Wiseman, Roger, 229, 230

yeast:
  cell cycles of, 192
  mismatch repair genes in, 167-
    170
  see also Kolodner, Richard
yeast artificial chromosomes
  (YACs), 220

Illustration Credits